Papermaking Science and Technology

a series of 19 books covering the latest technology and future trends

Book 18

Paper and Board Grades

Series editors
Johan Gullichsen, Helsinki University of Technology
Hannu Paulapuro, Helsinki University of Technology

Book editor
Hannu Paulapuro, Helsinki University of Technology

Series reviewer
Brian Attwood, St. Anne's Paper and Paperboard Developments, Ltd.

Book reviewer
Gary Smook, Writer/Consultant

Published in cooperation with the Finnish Paper Engineers' Association and TAPPI

Cover photo: Luonnonkuva-arkisto

ISBN 952-5216-00-4 (the series)
ISBN 952-5216-18-7 (book 18)

Published by Fapet Oy
(Fapet Oy, PO BOX 146, FIN-00171 HELSINKI, FINLAND)

Copyright © 2000 by Fapet Oy. All rights reserved.

Printed by Gummerus Printing, Jyväskylä, Finland 2000

 Printed on LumiMatt 100 g/m^2, Enso Fine Papers Oy, Imatra Mills

Certain figures in this publication have been reprinted by permission of TAPPI.

Foreword

Johan Gullichsen and Hannu Paulapuro

PAPERMAKING SCIENCE AND TECHNOLOGY

Papermaking is a vast, multidisciplinary technology that has expanded tremendously in recent years. Significant advances have been made in all areas of papermaking, including raw materials, production technology, process control and end products. The complexity of the processes, the scale of operation and production speeds leave little room for error or malfunction. Modern papermaking would not be possible without a proper command of a great variety of technologies, in particular advanced process control and diagnostic methods. Not only has the technology progressed and new technology emerged, but our understanding of the fundamentals of unit processes, raw materials and product properties has also deepened considerably. The variations in the industry's heterogeneous raw materials, and the sophistication of pulping and papermaking processes require a profound understanding of the mechanisms involved. Paper and board products are complex in structure and contain many different components. The requirements placed on the way these products perform are wide, varied and often conflicting. Those involved in product development will continue to need a profound understanding of the chemistry and physics of both raw materials and product structures.

Paper has played a vital role in the cultural development of mankind. It still has a key role in communication and is needed in many other areas of our society. There is no doubt that it will continue to have an important place in the future. Paper must, however, maintain its competitiveness through continuous product development in order to meet the ever-increasing demands on its performance. It must also be produced economically by environment-friendly processes with the minimum use of resources. To meet these challenges, everyone working in this field must seek solutions by applying the basic sciences of engineering and economics in an integrated, multidisciplinary way.

The Finnish Paper Engineers' Association has previously published textbooks and handbooks on pulping and papermaking. The last edition appeared in the early 80's. There is now a clear need for a new series of books. It was felt that the new series should provide more comprehensive coverage of all aspects of papermaking science and technology. Also, that it should meet the need for an academic-level textbook and at the same time serve as a handbook for production and management people working in this field. The result is this series of 19 volumes, which is also available as a CD-ROM.

When the decision was made to publish the series in English, it was natural to seek the assistance of an international organization in this field. TAPPI was the obvious partner as it is very active in publishing books and other educational material on pulping and papermaking. TAPPI immediately understood the significance of the suggested new series, and readily agreed to assist. As most of the contributors to the series are Finnish, TAPPI provided North American reviewers for each volume in the series. Mr. Brian Attwood was appointed overall reviewer for the series as a whole. His input is gratefully acknowledged. We thank TAPPI and its representatives for their valuable contribution throughout the project. Thanks are also due to all TAPPI-appointed reviewers, whose work has been invaluable in finalizing the text and in maintaining a high standard throughout the series.

A project like this could never have succeeded without contributors of the very highest standard. Their motivation, enthusiasm and the ability to produce the necessary material in a reasonable time has made our work both easy and enjoyable. We have also learnt a lot in our "own field" by reading the excellent manuscripts for these books.

We also wish to thank FAPET (Finnish American Paper Engineers' Textbook), which is handling the entire project. We are especially obliged to Ms. Mari Barck, the project coordinator. Her devotion, patience and hard work have been instrumental in getting the project completed on schedule.

Finally, we wish to thank the following companies for their financial support:

A. Ahlstrom Corporation
Stora Enso Oyj
Kemira Oy
Metsä-Serla Corporation
Metso Oyj
Raisio Chemicals Ltd
Tamfelt Corporation
UPM-Kymmene Corporation

We are confident that this series of books will find its way into the hands of numerous students, paper engineers, production and mill managers and even professors. For those who prefer the use of electronic media, the CD-ROM form will provide all that is contained in the printed version. We anticipate they will soon make paper copies of most of the material.

List of Contributors

Haarla, Ainomaija, M.Sc. (Tech.), MBA, Vice President, Corporate Marketing, Metso Corporation

Kimari, Outi, M.Sc (Econ.), Freelance Editor

Kiviranta, Ari, D.Sc. (Tech.), Process Development Manager, Metsä-Serla Oyj

Meinander, Paul Olof, M. Sc. (Tech), President, POM Technology Oy Ab

Niemi, Tapio, M. Sc. (Tech), R & D Director, Walkisoft

Paulapuro, Hannu, D.Sc. (Tech.), Professor, Department of Forest Products Technology, Paper Technology, Helsinki University of Technology

Preface
Hannu Paulapuro

Knowledge of different paper and board grades is important for everybody working in the field of forest products or related industry. It is needed in several practical situations, for example, in market analyses, in evaluating the potential of a given raw material, in rebuilding the existing production line, in developing the properties of a certain paper grade, and in compiling trade statistics. In order to serve these purposes, the paper and board grade information should include at least the basis of classification, end-use description and requirements, raw material application, and the main features of the manufacturing technology.

There is no generally accepted or standardized paper grade classification system. Different classification systems have been developed and used by various international organizations, like FAO and CEPI, market associations, consultants, and the forest industry companies themselves. Different bases have been used in classification: raw material compositions, end uses, basis weights, and manufacturing technologies. Even in the same classification system different bases are used when going into the finer subdivision of the grades. Furthermore, there are differences in the classification of paper and board grades between geographical areas, like Western Europe, North America, and Japan.

Paper and board grade classification is also made complex by the continuous development of the grades. This complexity has been accelerating during recent years for several reasons. Changes both in the markets and in the technical possibilities have been behind the grade development. Emergence of new paper grades has been forced by the development of new printing and copying methods. Raw material basis has widened due to the application of new fillers and pigments, new types of pulps, and wider use of recycled fibers. Papermaking technology has also advanced rapidly, especially in the field of coating and surface treatment, making it possible to produce more tailor-made products to satisfy the increasing customer needs. A recent publication presented an interesting new view of the evolution of paper grades[1].

As a result of this development, the definitions of many paper grades based on conventional classification systems have been blurred and significant overlapping exists. A good example is the situation between the traditional woodcontaining and woodfree paper grades. Earlier these paper grades had clearly distinctive application areas. Currently many of these grades compete in the same end use, with the decisive factor being the quality-cost relationship. It also seems to be a more general trend in the evolution of the paper and board grades that the final classification basis is the end use and its requirements, irrespective of the raw materials and manufacturing technology used.

There are hundreds of different types of papers and boards produced in the world and new types are continuously emerging. This book does not aim to examine them all, but covers the most important grades in the international markets. The classification is based on that used by the European forest industry companies, with the main division being:

- Printing and writing papers
- Paperboards
- Tissue
- Specialty papers.

Each group is divided into subgroups and main grades and the grades are examined in view of the end uses, raw material composition, and manufacturing technology. Air-laid papers are dealt with in a separate group. Data on the end-use properties are not normally presented, since they are changing continuously due to ever-increasing quality competition.

Some comparisons are made with the other internationally used classification systems, although the text does not aim to be complete in this respect.

Finaly, I would like to thank our contributors for the well-done work. Their expertise, experience and insight into the subject made the editor's task both pleasant and rewarding.

Otaniemi, February 2000

Hannu Paulapuro

References

1. Vasara, P., Jallinoja, K., Lobbas, P., Luutiala, P., Paperi ja Puu 81(1):34(1999).

Table of Contents

	List of Abbreviations	11
1.	Printing and writing papers	14
2.	Paperboard grades	55
3.	Tissue	75
4.	Air-laid paper	95
5.	Specialty papers	101
	Conversion factors	131
	Index	132

List of Abbreviations

AFPA	American Forest Products Association
CAD	Computer aided design
CB	Coated back
CD	Cross direction
CEPI	Confederation of European Paper Industries
CF	Coated front
CFB	Coated front and back
CMC	Carboxy methyl cellulose
CSF	Canadian standard freeness
CTMP	Chemithermomechanical pulp
DIP	Deinked pulp
ECF	Elemental chlorine free
ECT	Edge crush test
EU	European union
EVA	Economic value added
FAO	Food and Agriculture Organization of the United Nations
FBB	Folding boxboard
FCO paper	Film coated offset paper
FCT	Flat crush test
FDA	Food and Drug Administration
GDP	Gross domestic product
GW	Groudwood pulp
HF Score	Hand feel score
HSWO	Heat set web offset
HW	Hardwood
HWC papers	High weight coated papers
LB	Latex bonded
LBAL	Latex bonded air laid paper
LDTA	Long dwell time applicator
LPB	Liquid packaging board
LWC papers	Light weight coated papers
LWCO papers	Light weight coated offset papers
LWCR papers	Light weight coated rotogravure papers
MBAL	Multibonded air laid paper
MD	Machine direction
MF	Machine finished
MFC papers	Machine finished coated papers

MFS papers	Machine finished special newsprint papers
MG	Machine glazed
MIT	Massachusetts Institute of Technology
MWC papers	Medium weight coated papers
NCR	No carbon required
OCC	Old corrugated containers
OCR	Optical character reading
OEM	Office equipment manufacturer
PAA	Polyacryl amide
PAT	Pin adhesion test
PE	Polyetylene
PGW	Pressure groudwood pulp
PIRA	Paper Industry Research Association
PM	Paper machine
POS	Point of sale
PPS	Parker print surf
PVA	Polyvinyl alcohol
RCF	Recycled fiber
SAP	Superabsorbent granules
SB	Styrene butadiene
SBS	Solid bleached sulphate
SC Cat	Supercalendered paper for catalogs
SC papers	Supercalendered papers
SC-A+, SC-A, SC-B, SC-C	Subgrades of the product group SC papers
SCO papers	Supercalendered paper for offset printing
SCR papers	Supercalendered papers for rotogravure printing
SCT	Short column test
SDTA	Short dwell time applicator
SEC	U.S. Securities and Exchange Commision
SUS	Solid unbleached sulphate
SW	Softwood
TAD	Through-air drying
TBAL	Thermobonded air-laid paper
TCF	Total chlorine free
TD papers	Telephone directory papers
TMP	Thermomechanical pulp
ULWC papers	Ultralight weight coated papers
UV radiation	Ultraviolet radiation
WF papers	Chemical pulp dominating paper grades (also called fine papers or woodfree papers)
WFC papers	Woodfree coated papers (also called coated fine papers or coated freesheet)
WFU papers	Woodfree uncoated papers (also called uncoated fine papers or uncoated freesheet)
WLC	White lined chipboard

CHAPTER 1

Printing and writing papers

1	**Business environment**	**14**
1.1	Global demand and supply	15
1.2	Industry structure and trends	18
1.3	Scale of operations	20
2	**Products**	**20**
2.1	Mechanical pulp dominating paper grades	24
	2.1.1 Newsprint	24
	2.1.2 SC papers	28
	2.1.3 Coated mechanical papers	30
2.2	Chemical pulp dominating paper grades	35
	2.2.1 Uncoated fine papers	36
	2.2.2 Coated fine papers	38
	2.2.3 Special fine papers	39
2.3	Printing and writing paper grades: general trends	41
3	**Different regional paper grade classifications**	**42**
3.1	European paper grade classification	42
3.2	American paper grade classification	43
3.3	Japanese paper grade classification	43
3.4	Printing and writing paper grade classification according to FAO	45
3.5	Printing and writing paper classification according to CEPI	46
3.6	Some comments on different classification systems	46
4	**Some aspects on choosing a paper grade for a specific purpose**	**46**
5	**Key drivers for substitution between printing and writing paper grades**	**49**
6	**Some comments on the challenge from electronic media**	**49**
	References	52

CHAPTER 1

Ainomaija Haarla

Printing and writing papers

1 Business environment

Printing and writing papers are those paper grades which are used for newspapers, magazines, catalogs, books, commercial printing, copying, business forms, and stationery as well as for laser and digital printing. They account for about 30% of the world's paper and board markets. The estimated value of the segment was about US$ 65 billion in 1997. Per capita consumption of printing and writing papers varies a lot regionally as Fig. 1 indicates. Consumption of these paper grades is concentrated in North America, Western Europe, and Japan. In North America it is 106 kg per capita per annum, in Japan 92 kg per capita per annum, and in Western Europe 64 kg per capita per annum. On an average, it is about 18 kg per capita per annum in the world. The regional variations stem from different markets and differently developed end use markets and also from the supporting and surrounding business environment.

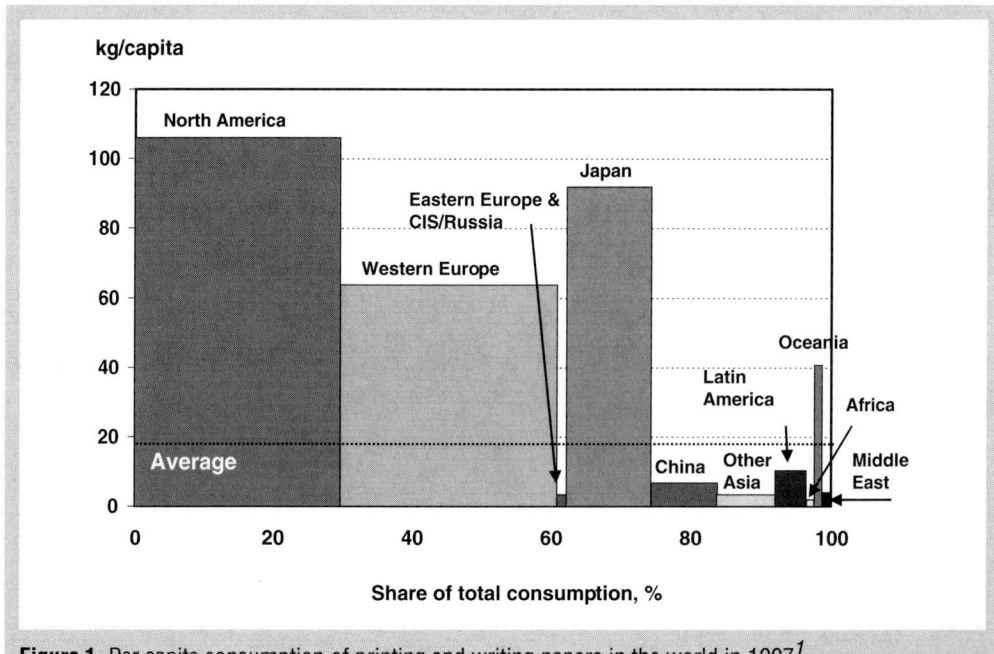

Figure 1. Per capita consumption of printing and writing papers in the world in 1997[1].

Printing and writing papers

The biggest challenges facing the printing and writing paper industry come from not only efficient use of capital and human and raw material resources, but also from proper response to the challenges set by the emerging electronic media as well as environmental demands as shown in Table 1.

Table 1. Challenges facing the printing and writing paper industry.

Low building costs of new capacity per product unit	Mixed use of various paper grades within one end use
	Branding
Low fiber cost per product unit	Newcomers into the industry especially in Asia
Low energy consumption per product unit	Cyclicality
High production efficiencies	Challenge from an electronic media
Supplying global markets from transnational manufacturing	Increasing consumer awareness in terms of environmental matters[2]
Timely service and logistics	Competition on intellectual capital

1.1 Global demand and supply

Global demand for printing and writing papers was at the level of 99 million t in 1998. The demand is divided between the main regions and product groups as Fig. 2 indicates.

General economic condition is the main driving force for paper demand. Overall consumption of printing and writing paper grades is often compared with the development of GDP as shown in Fig. 3. There have recently been some signals of a weakening correlation.

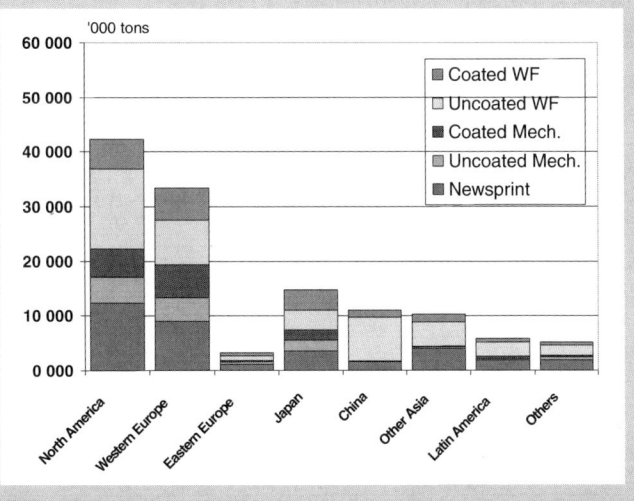

Figure 2. Printing and writing paper demand by region and by product group 1997[1].

Demand growth has been good and is evaluated to stay at a 3%–4% level for the next 10–15 years. This is based on a scenario where print and electronic media support each other's growth.

CHAPTER 1

Advertizing activity is followed closely as an indicator for future newsprint and magazine paper consumption. New laser and digital printer installations are regarded as one indicator for future fine paper demand.

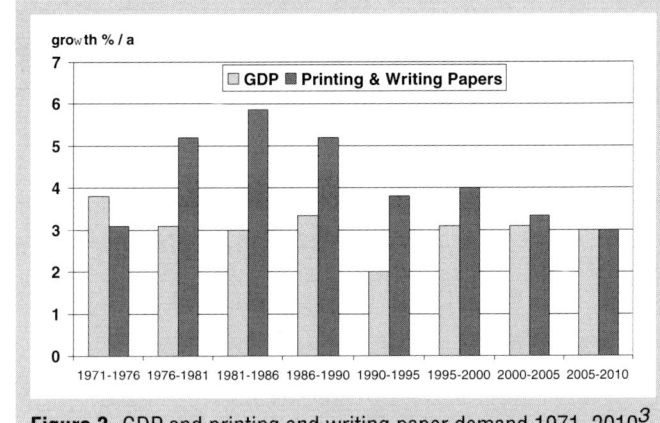

Figure 3. GDP and printing and writing paper demand 1971–2010[3].

Demand for *mechanical pulp dominating paper* grades, such as newsprint, SC, and coated mechanical papers, is growing somewhat slower than that for *chemical pulp dominating papers,* such as uncoated and coated fine papers. Expected world average demand growth for newsprint is 1.8%, for uncoated mechanical papers 2.1%, for coated mechanical papers 4.0%, for uncoated fine papers 3.6%, and for coated fine papers by 4.7% between 1996 and 2005 per annum according to Jaakko Pöyry Consulting (update July 1999).

Fine papers (woodfree papers) dominate in Asia, whereas the relative demand of newsprint remains the same regardless of a region as is to be seen in Fig. 4. Mechanical pulp dominating magazine papers are mostly produced in Europe as shown in Fig. 5, where suitable raw materials exist.

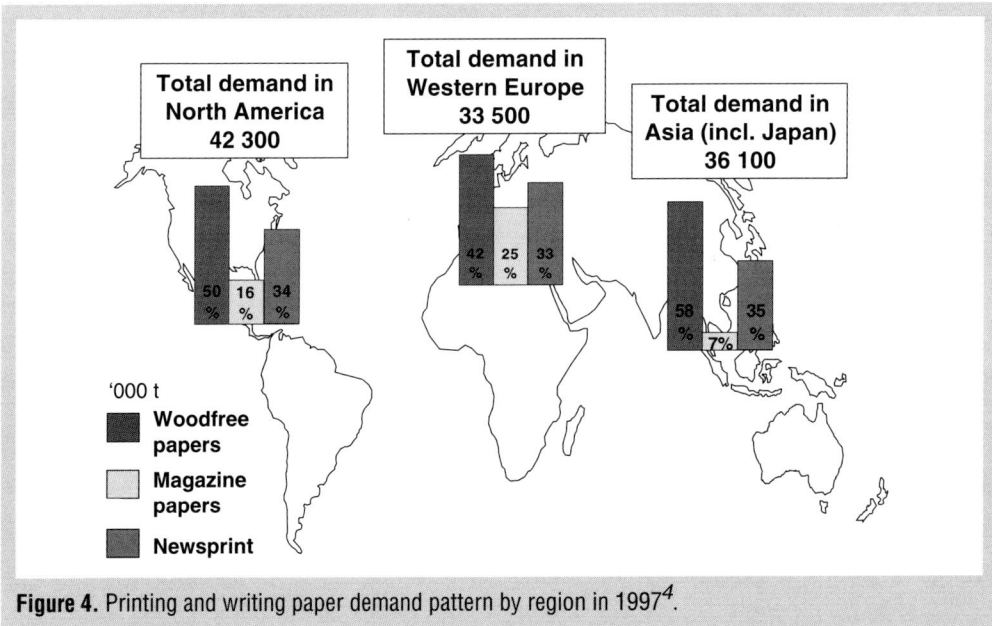

Figure 4. Printing and writing paper demand pattern by region in 1997[4].

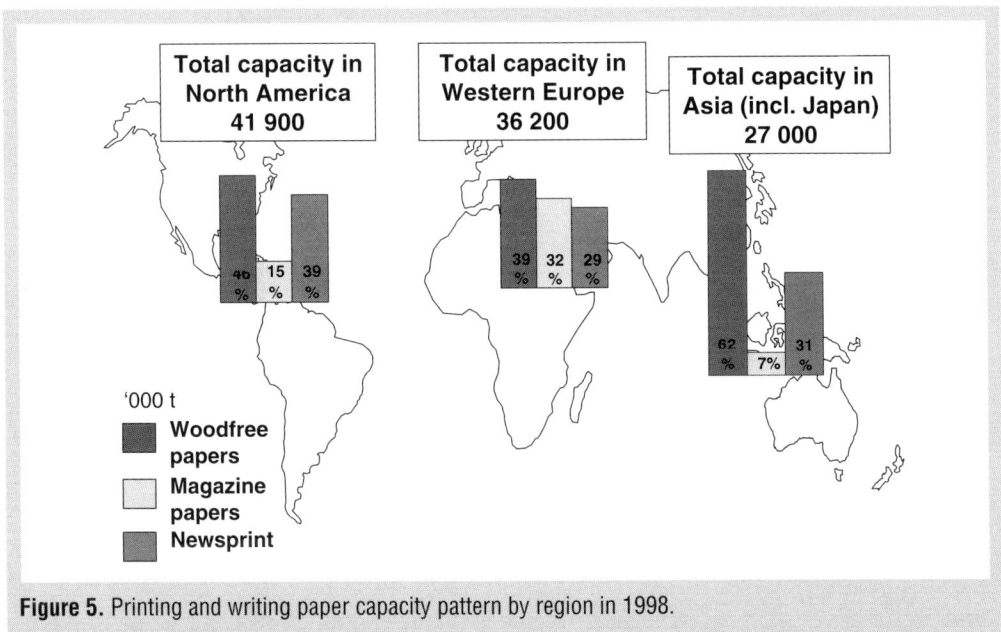

Figure 5. Printing and writing paper capacity pattern by region in 1998.

Printing and writing paper flows are global. Those flows differ, however, by grade. Mechanical pulp dominating paper grades are currently traded mostly from the Northern hemisphere to the Southern hemisphere. They are based on Northern spruce fiber for a major part. Some 55% of uncoated mechanical pulp dominating papers' consumption is traded across country borders and also 45% of coated mechanical pulp dominating papers. However, chemical pulp dominating papers, also known as fine papers, are currently produced in proximity to the markets. Their production is based on fast-growing hardwood. Only 17% of uncoated fine papers are traded internationally and about 30% of coated fine papers. The situation is rapidly changing as the new Asian producers have access to economical fiber. This may lead to a turnaround in fine paper flows in the future. The future of real fine paper

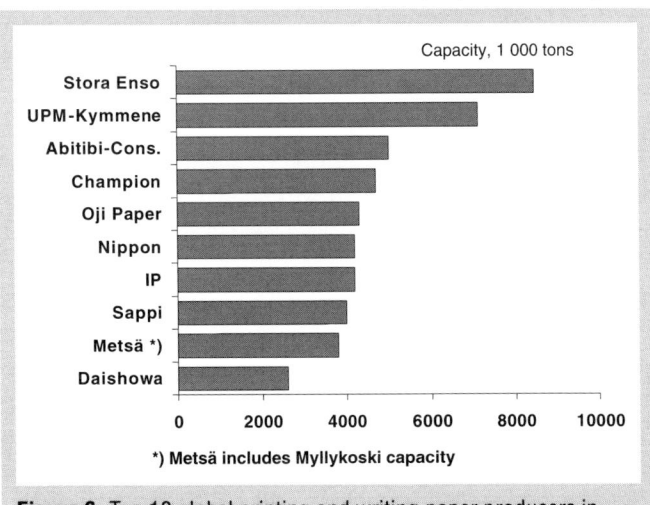

Figure 6. Top 10 global printing and writing paper producers in summer 1998[5].

CHAPTER 1

demand in China plays a key role. The existing swing capacity, especially in fine papers, might also alter the balance.

The world's largest printing and writing paper producers come from Scandinavia, North America, and Japan. Figure 6 lists the world's largest global printing and writing paper producers in the summer of 1998.

Printing and writing paper supply is dynamically changing through startups of new greenfield or brownfield machines in Asia and in Europe, in particular, through rebuilds and shutdowns. A total figure for existing printing and writing machinery amounted to 2 550 in 1997. Figure 7 shows the development of the number of all new paper machines, rebuilds, and shutdowns in the whole world since 1988.

The shutdowns peaked in the early 1990s with more than 200 paper machines per year. Existing machinery is rebuilt at a level of 130 machines per year. A number of new paper machines has stayed at the level of 70–80 paper machines per year during the recent years. Capacity change is, however, what matters most. Capacity increase through new paper machines was 3 to 4 times as large as capacity decrease through shutdowns in 1997 and 1998. The current tendency is to buy existing capacity rather than to build it.

1.2 Industry structure and trends

Paper industry is very fragmented compared to other industries. The top 10 producers account for only 20%; whereas, for example, in the chemical industry this figure is 40%. The global capacity share of the biggest producer, International Paper, was only 3.1% in 1997. A clear tendency among printing and writing paper producers is toward bigger global companies through mergers, acquisitions, and various alliances which will have

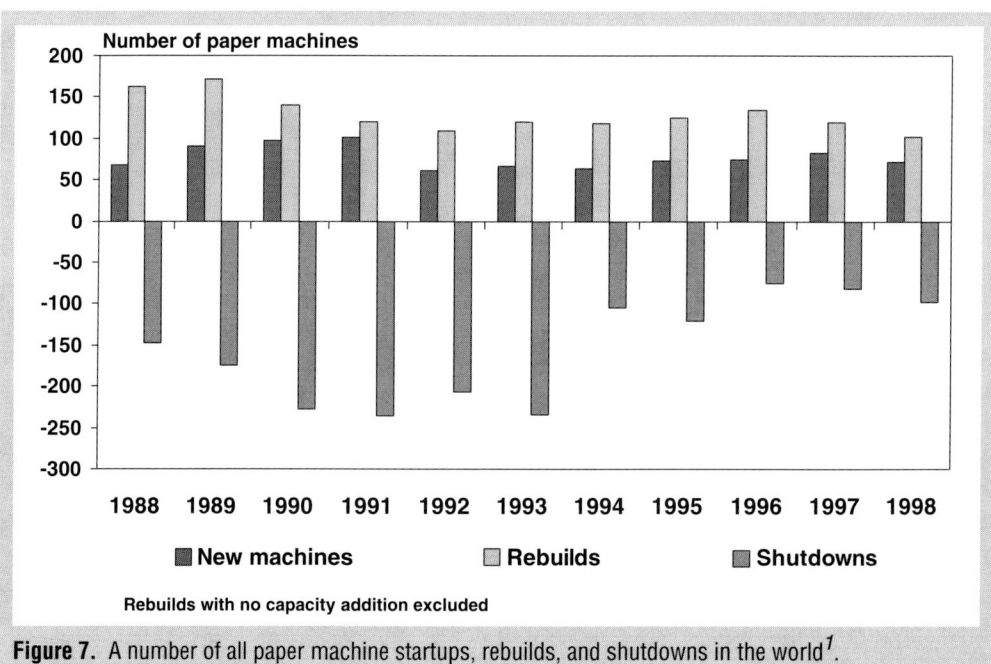

Figure 7. A number of all paper machine startups, rebuilds, and shutdowns in the world[1].

local presence. Global companies are serving global markets as well as customers and will be doing more so through transnational manufacturing[6]. Owing to the recent Asian economic turmoil, new alliances between Asian, European, North American, and Japanese producers are currently dynamically shaping the global paper industry structure further.

Consolidation of the global paper and board industry started in the late 1980s as Fig. 8 indicates. Consolidation varies by printing and writing paper product group as shown in Fig. 9. Knowledge exchange plays an increasingly important role, for example, in the areas of technology and product development in the period of consolidation. The following reasons for increased mergers and acquisitions activity have been given in another volume of this book series, *Economics of Pulp and Paper Industry*:

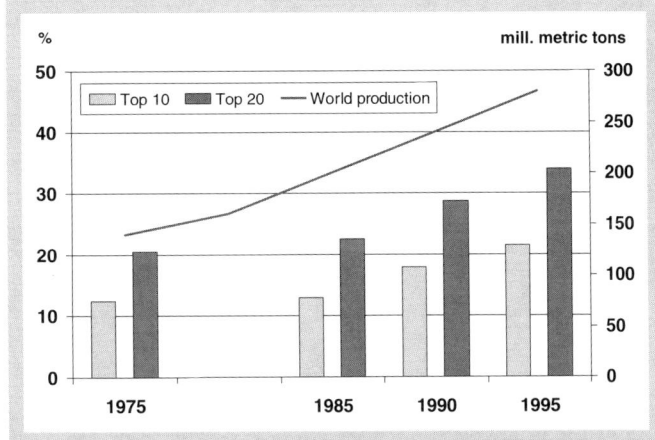

Figure 8. Consolidation of paper and board industry since the mid-1970s.

	C-1	C-5	C-10	C-15
Newsprint	8 %	38 %	67 %	88%
Uncoated mech.	18 %	54 %	80 %	96 %
Coated mech.	17 %	52 %	79 %	95 %
Uncoated WF	20 %	56 %	83 %	96 %
Coated WF	14 %	49 %	77 %	96 %

C-1 = The share of the biggest producer among top 25
C-5 = The share of top 5 producers among top 25
C-10 = The share of top 10 producers among top 25
C-15 = The share of top 15 producers among top 25

Figure 9. Consolidation in printing and writing product group[1].

- The growth of importance of recycled fiber as a raw material for the paper industry; Scandinavian producers' limited access to this raw material has forced them to merge with central European companies close to vast resources.
- The importance of economy of scale in a commodity business such as paper industry
- Cyclicality of product prices and decreased predictability; a need to lower the risk regarding the balance sheet
- A desire to strengthen a market position without a new capacity investment
- The concentration of customer base, newsprint, and magazine publishers in particular.

CHAPTER 1

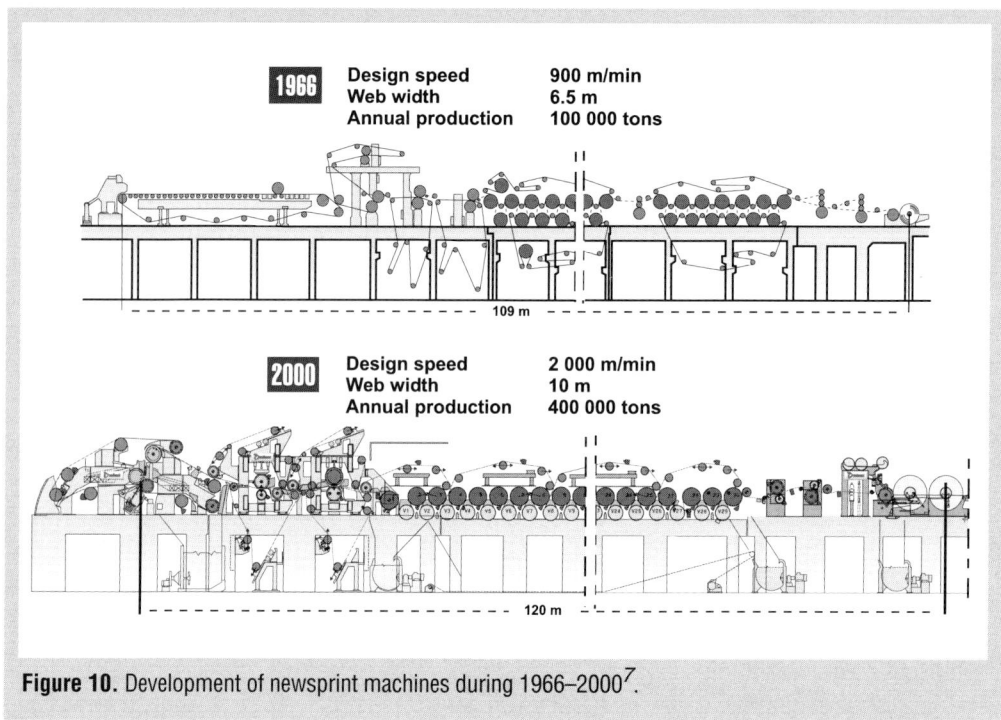

Figure 10. Development of newsprint machines during 1966–2000[7].

Increased profitability demands set by shareholders and developed financial management tools such as Economic Value Added (EVA) have also speeded up consolidation development.

1.3 Scale of operations

Today the main part of printing and writing papers is produced with wide and fast machines. Figure 10 illustrates the development using newsprint paper machine as an example. One trend is to cut off the total length of a paper machine. Simultaneously paper machine speed is increased and production efficiency is improved to raise productivity. Some smaller scale mill concepts, urban mini-mills, are currently under development also for printing and writing papers. Feasability is based on smaller investment costs, flexible manufacturing process, more economical logistic costs, and availability of recycled fiber from the neighborhood.

2 Products

Printing and writing papers are divided into two groups, *mechanical pulp dominating paper grades* and *chemical pulp dominating paper grades*. Figure 11 shows the printing papers' range used in this chapter. The key classification criterion is thus the nature of the main raw material, pulp. Recycled fiber (RCF) is currently also one important raw material. It is used when its availability is secured and when its usage is economically sound. There are regional differences in availability due to differences in population density and in the organization of collection. The usage of RCF is commented upon later when dealing with various paper grades where relevant.

Printing and writing papers

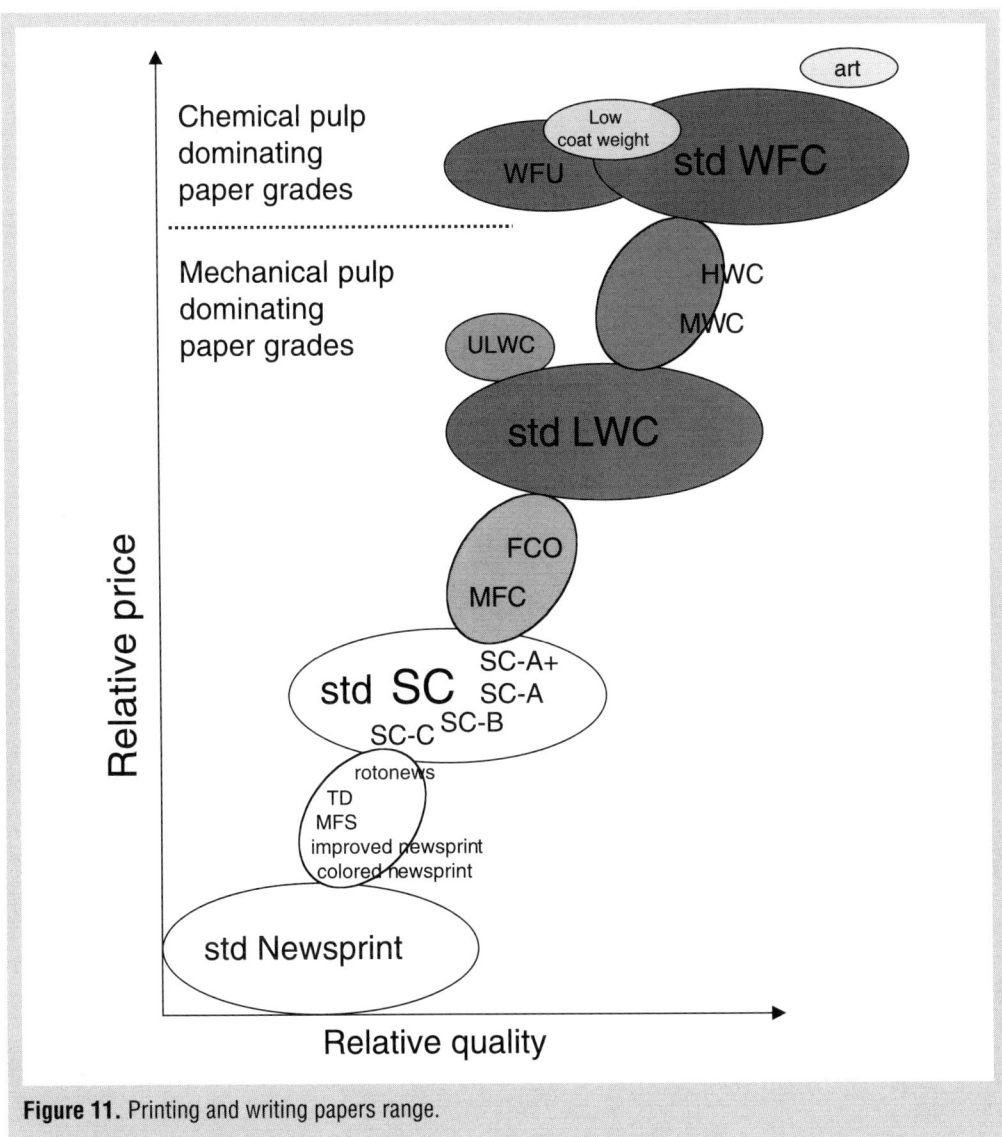

Figure 11. Printing and writing papers range.

One of the shortcomings of the classification now in use is its production orientation. It does not indicate, for example, in which applications the paper grade in question would give the maximum value to a customer. Many paper producers have sharpened their customer-focused approach recently. So, for example, end use (such as magazine) or end user (such as magazine publisher) -based classifications could be more relevant from a business decision maker's perspective. However, there is currently very limited public data available and, in addition, the standards are missing. Table 2 illustrates typical end uses and the most common competing paper grades for those papers discussed in this chapter.

Table 2. Typical end uses and the most common competing paper grades for printing and writing papers.

	Paper grade	Typical end uses	The most common competing paper grades
MECHANICAL PULP DOMINATING PAPER GRADES			
Newsprint			
	Standard newsprint	newspapers newspaper supplements inserts and flyers	special newsprint papers (MFS product group)
Special newsprint grades	Improved newsprint	newspapers newspaper supplements inserts and flyers	SC-C rotonews
Special newsprint grades	TD papers	telephone directories other directories timetables	standard lightweight newsprint
Special newsprint grades	Colored newsprint	same as for standard newsprint	–
Special newsprint grades	Rotonews (mainly in North America)	newspaper supplements commercial printing low-end catalog	SC-C MFS papers
Special newsprint grades	MFS-papers (a wide range of uncoated mechanical pulp dominating papers)	newsprint supplements newspapers freesheets inserts magazines direct mail bulky grades for pocket and comic books	competing paper grades depend on end use
SC papers			
	SC-A+, SC-A, standard SC	rotogravure papers for magazines, catalogs, commercial printing offset papers for TV-listings, magazines, direct mail, supplements	ULWC FCO
	SC-B	low-end magazines newspaper supplements low-end catalogs	standard SC special newsprint papers
	SC-C	same as for standard newsprint and SC-B	improved newsprint rotonews SC-B
Coated mechanical papers			
	Standard LWC	magazines catalogs inserts commercial printing	WFC FCO SC-A+ depending on end use
	ULWC	rotogravure printed catalogs in Europe, offset printed magazines in the U.S.	SC-A

Table 2. Typical end uses and the most common competing paper grades for printing and writing papers. (Continued)

Paper grade	Typical end uses	The most common competing paper grades
MECHANICAL PULP DOMINATING PAPER GRADES		
MWC	high-end special interest magazines catalogs direct mail other advertising	WFC
HWC	high-end special interest magazines catalogs magazine covers direct advertising	WFC
MFC	special interest magazines catalogs commercial printing books	LWCO SCO MFS
FCO	special interest magazines catalogs	LWCO
CHEMICAL PULP DOMINATING PAPER GRADES		
Uncoated fine papers		
Standard WFU	cut-size business forms envelopes direct mail books and manuals	competing paper grades depend on end use
Offset papers (a large variety of machine finished papers)	commercial printing books magazines catalogs	–
Lightweight papers (low-weight offset papers)	direct marketing bibles dictionaries	–
Coated fine papers		
Standard WFC (a large variety of coated chemical pulp dominating papers)	magazines (dominant in Europe) catalogs (dominant in the U.S.) direct mail books and manuals labels	high-bright LWC-, MWC- and HWC-papers art papers
Low coat weight papers	books directories timetables brochures	–
Art papers	THE MOST DEMANDING END USES such as illustrated books, calendars, brochures	

Table 2. Typical end uses and the most common competing paper grades for printing and writing papers. (Continued)

Paper grade	Typical end uses	The most common competing paper grades
CHEMICAL PULP DOMINATING PAPER GRADES		
Special fine papers		
Copy papers	copying non-impact printing	–
Digital printing papers (expanding variety of uncoated and coated fine papers)	manuals price lists direct mail low-volume paperbacks and hard-cover books	competing paper grades depend on end use
Continuous stationary	listings custom-made forms	–

2.1 Mechanical pulp dominating paper grades

Mechanical pulp dominating paper grades, also known as wood-containing papers or mechanical printing papers, comprise various newsprint grades, SC paper grades, and coated mechanical paper grades. They contain 25%–100% mechanical pulp depending on the paper grade – but usually more than 50%. Coarseness of the mechanical pulp varies from newsprint's CSF 95 ±10 ml to coated mechanical's and standard SC's CSF 35 ±10 ml. Chemical pulp is added to increase strength and to guarantee both paper machine, coater, and printing machine runnability. Minerals used as fillers or in the coating are used to improve printability through improved smoothness and gloss and also to increase brightness and maintain opacity.

The main benefits of papers containing mechanical pulp are good opacity, even at low basis weights, and good printability. These papers are also an economical option for a publisher or a printer especially in mass-publications, because lower basis weight can be used and because the price of mechanical pulp is lower than the price of chemical pulp. The disadvantage of mechanical pulp dominating paper is its tendency to yellow when exposed to UV radiation, which thus limits archiving properties. Mechanical pulp dominating papers are used in such end use applications where the life cycle of printed products is limited. Coated mechanical pulp dominating papers can be stored longer than uncoated ones.

2.1.1 Newsprint

Product group newsprint comprises standard newsprint and special newsprint grades such as telephone directory paper, colored newsprint, rotonews, and a large group of machine finished specialty papers, MFS papers including, for example, book papers. Newsprint paper grades are delivered only in reels.

Standard newsprint grades

Standard newsprint grades are mainly used for printing newspapers, but also for newspaper supplements, weeklies, and less demanding commercial printing. Figure 12 shows typical end uses for standard newsprint in Western Europe, and Fig. 13 shows those in the United States. The main paper grade competition comes from special newsprint grades.

The most common basis weights are 45 and 42.5 g/m^2, but also basis weights of 40 and 48.8 g/m^2 are commonly used. There has been a tendency to reduce the basis weight of newsprint to achieve lower transport, handling, and storage costs. The increasing use of recycled fiber, however, has slowed down this development. Because of its better runnability, 45 g/m^2 paper is often an option when working in the printing house under time pressure. A light, 35 g/m^2, or even lighter newsprint is produced for air mail editions. Increasing four-color printing in certain cases can restrict a reduction in basis weight due to strike through.

The most important properties for standard newsprint are density, brightness and color, smoothness of surface, oil absorption, and strength properties (tensile and

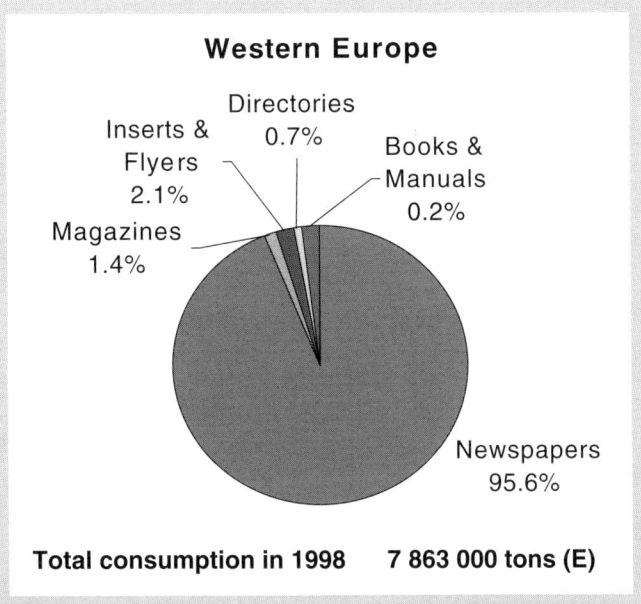

Figure 12. Typical end uses for standard newsprint in Western Europe in 1998[8], (E = estimate).

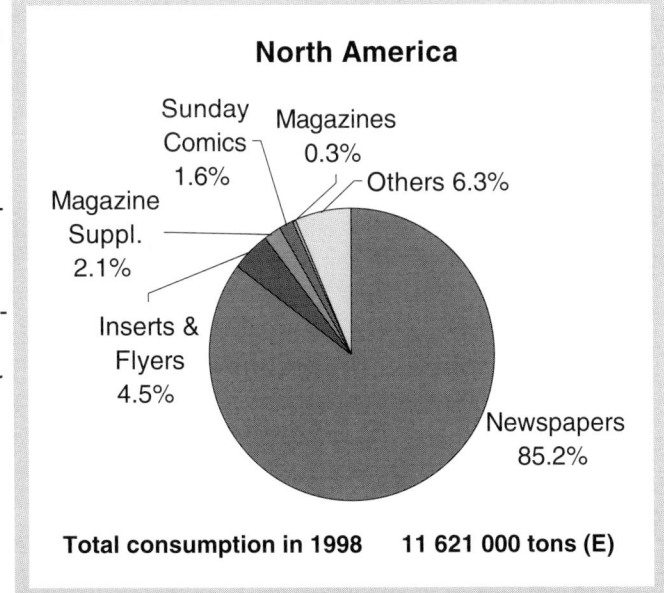

Figure 13. Typical end uses for standard newsprint in North America in 1998[9].

CHAPTER 1

tear). Runnability at a printing press is of utmost importance due to strict production schedules. European and U.S. shades for standard newsprint differ somewhat from one another, with newsprint used in the United States being slightly redder and in Europe slightly greener.

The quality of newsprint varies a lot between producers. The essential difference comes from the main raw material, which is either recycled or virgin fiber. Newsprint produced from different raw materials behaves in a different manner at a printing press and is not interchangeable without additional adjustments made at the printing press.

Standard newsprint grades can contain up to 100% recycled fibers, which is the case in Central Europe for instance, or up to 100% mechanical pulp. Mechanical pulp is either groundwood, TMP, PGW, or CTMP. The use of chemical pulp is rare in Europe, but outside Europe it can be as much as 30%. The use of RCF can bring filler content above 15% in extreme cases. Standard newsprint is mainly produced with the most modern high-speed paper machines in a cost competitive manner. Gap former technology is applied to the newest paper machine installations. A deinking plant is often one part of a new standard newsprint machine line. Standard newsprint is calendered on-line. Soft calendering with high temperature and linear loads is becoming more common in newsprint production. Higher strength at the same smoothness level and smaller linting tendency can then be achieved.

The main printing methods for standard newsprint are cold set web offset, flexo, and letterpress. Offset and flexo printing have replaced letterpress in many cases.

Special newsprint grades

There is an increasing number of special newsprint grades. The following text comments on improved newsprint grades and separately telephone directory paper, colored newsprint, rotonews, and a broad group of machine finished specialties (MFS papers) with emphasis on the main differences as compared with standard newsprint.

Improved newsprint

Improved newsprint grades are used for many different purposes such as in upmarket newspapers, newspaper reprints, newspaper supplements, weeklies, and journals as well as comic and pocket books. Figure 14 illustrates typical end uses for improved newsprint in Western Europe. Main competing paper grades are SC-C and rotonews.

Improved newsprint grades cover a wide basis

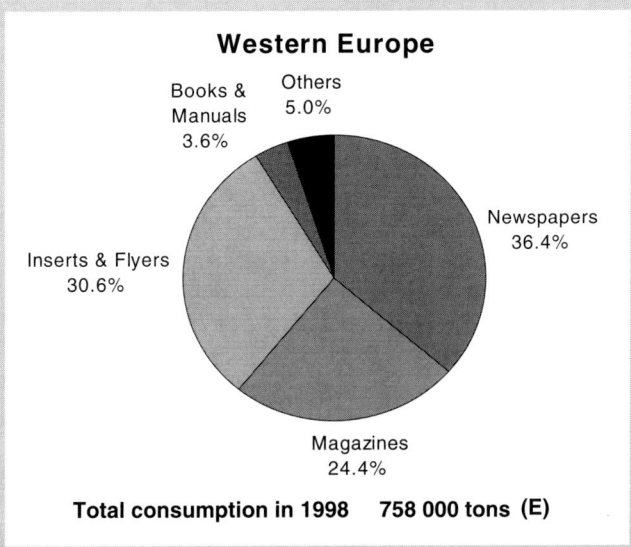

Figure 14. Typical end uses for improved newsprint in Western Europe in 1998[8].

Printing and writing papers

weight area from 36 g/m² to 70 g/m², the main area being from 52 to 55 g/m². In many cases, improved newsprint grades are heavier, brighter, and stiffer than standard newsprint. Chemical pulp content can also be higher than for standard newsprint. Mechanical pulp is specially bleached for high bright grades. Where recycled fiber is used, bleaching is also necessary. Recycled fiber content can vary from 40% to 50% and filler content from 0% to 8%. Some papers belonging to this product group can also be slightly supercalendered and also colored.

Improved newsprint is often produced by older newsprint machines, which were originally designed to produce standard newsprint, but were not competitive enough to continue in that sector. Existing production technologies vary a lot. Machine structure, availability of raw materials, and also market potential define the nature of these niche products.

Improved newsprint is mainly printed in offset and flexo.

Telephone directory paper

Telephone directory papers (TD papers) are lightweight newsprint paper grades. In addition to telephone directories, these papers are used for other directories and timetables. TD papers compete with standard lightweight newsprint in the markets.

TD papers can be white or colored, for example, yellow or pink. TD papers are often "tailor-made" products according to the final customers' needs, which, in many cases, is a telephone company.

The basis weight varies from 28 to 40 g/m².

Unlike standard newsprint, TD papers are mainly made of virgin fibers. Paper machines are more often old standard newsprint machines, which have been modernized to produce TD papers. In addition, soft calendering is gaining more ground here as a finishing method.

Colored newsprint

Technically this grade is close to standard newsprint and is used for the same end uses. The basis weight varies from 40 to 52 g/m². The most common colors are salmon-red, pink, yellow, blue, and green. Sometimes colored newsprint is classified as a paper grade of machine finished specialty grade. Recyclability problems of colored newsprint are not completely solved yet.

Rotonews

Rotonews is used for newspaper supplements, commercial printing, and also for low-end catalogs. It is mainly produced and consumed in North America with older standard newsprint machines, but it is also produced in Europe. Fiber furnish of this grade is as in standard newsprint, but pigment content can go slightly above the level of the latter. Basis weights start from 40 g/m² and normally go up to 55 g/m². Rotonews can also be supercalendered, and its surface smoothness is better than in standard newsprint. It is also slightly brighter than standard newsprint. Competing paper grades are SC-C paper and some machine-finished specialties.

CHAPTER 1

MF specialties
Machine-finished specialties (MFS papers) consist of a wide range of uncoated mechanical pulp dominating paper grades with varying end uses and properties. They are mainly used for newsprint supplements and newspapers, but also for freesheets, inserts, magazines, and direct mail. Bulky MFS papers are used for pocket books and comic books. These papers are often bulkier, heavier, and brighter than other uncoated mechanical pulp dominating papers. Customer-specific tailor-made products are also produced from MFS papers. Increased use of four-color printing is driving up the demand for improved newsprint grades with high brightness. Some surface treated paper grades have recently been introduced for cold set web offset printing to be used as supplements for example. MFS papers are often produced at paper machines that are not competitive enough to produce standard newsprint. In some cases, investment to improve bleaching has been necessary.

2.1.2 SC papers

SC papers, supercalendered papers, form a product group within magazine paper grades, where mechanical pulp dominates and which does not have any surface coating. Instead, they can contain up to 35% minerals as fillers. Paper quality is largely based on the quality of fibrous raw materials and fillers and their distribution in the sheet from top to bottom. For that reason, SC papers are among the most difficult products to manufacture within printing and writing paper grades. SC-A+, SC-A, SC-B, and SC-C are the subgrades within the product group SC papers. SC papers are sold in reels.

SC-A+, SC-A, standard SC

The main end uses for SC-A rotogravure grades are in magazines, catalogs (low basis weight SC papers in particular), and commercial printing. The use of SC papers in magazines, as well as in sales catalogs, has in recent years increased especially in the United States and is expected to rise further in the future due to a growth in the local production. Some European producers have even launched a special grade, SC Cat, for the sales catalogs printed in gravure. SC offset grade is used for TV listings, magazines, direct mail adver-

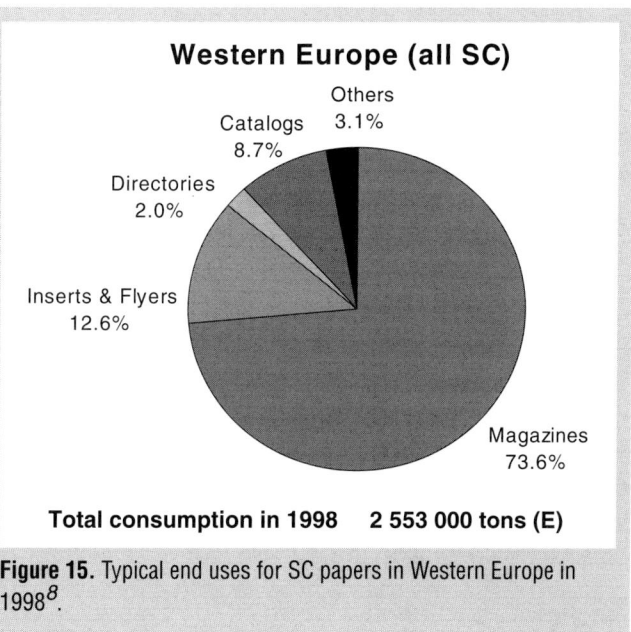

Figure 15. Typical end uses for SC papers in Western Europe in 1998[8].

tising, and also for Sunday supplements. Figure 15 shows typical end uses for SC papers in Western Europe and Fig. 16 shows for SC-A and SC-B in the United States. Rotogravure printing dominates: 80% of SC papers goes for rotogravure, 20% is used in offset.

Competing paper grades for top quality SC rotogravure papers are mainly ultralight weight coated papers and for SC offset, film coated offset. SC-A+ grade especially has taken some market share from coated mechanical pulp dominating papers due to a decreased quality difference. SC-A is a viable option when price and information capacity are important factors.

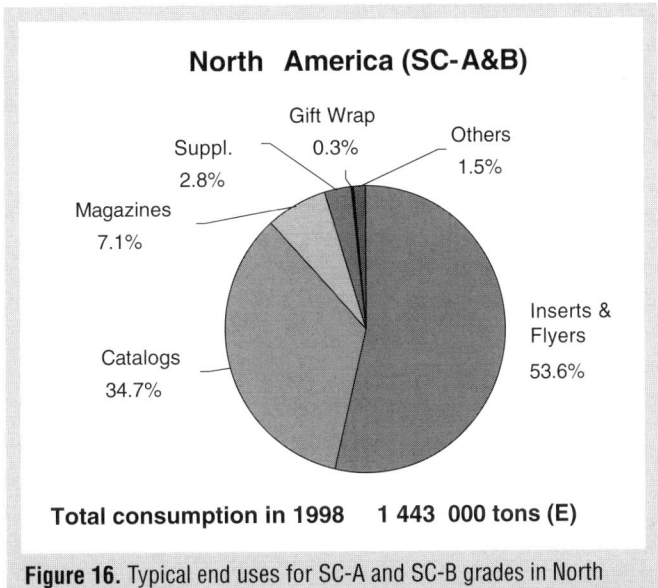

Figure 16. Typical end uses for SC-A and SC-B grades in North America in 1998[9].

Basis weights range from 39 to 80 g/m² with 52, 56, and 60 g/m² being the most typical. Due to high filler content, fine mechanical pulp, and heavy supercalendering, SC rotogravure paper is dense, smooth, and glossy enough even for a more demanding rotogravure printing. Brightness is typically on the level of 68%–70% (ISO) in 56 g/m² paper. Opacity is then 90%–91%. In offset grades, brightness can go up to 75%. Printed gloss is important in magazine end use where there are a lot of advertisements. Printing smoothness, good gloss, and opacity in low basis weights are emphasized in catalog end use due to vogue pictures. Stiffness and bulk are important properties in inserts and supplements due to a small number of pages. Rotogravure printers demand high production efficiency with high speeds (good dimensional stability is a necessity), low overall waste, and low absorption of printing ink.

For SC offset papers, the most important properties are good surface strengths (both dry and wet), good gloss (especially printed gloss), even setoff of the printing ink, dimension stability, heat resistance during drying, and low as well as even ink absorption.

SC papers are produced from 70%–90% conventional or pressurized groundwood (GW or PGW) or thermomechanical pulp (TMP), and 10%–30% chemical pulp. TMP has good strength properties, whereas PGW has good optical properties. Filler content can reach 35%–36% in 60 g/m² SC rotogravure paper and is lower in offset grades varying from 15% to 30%. The main filler is clay, but talc is also used in rotogravure grades. Additionally, some special pigments are used to increase opacity and brightness.

SC-A and SC-A+ grades are produced with the newest machines and the latest technology. The most common former types are either hybrid or gap formers. In addition, the press section often contains four nips to decrease two-sideness. Multilayering technology, additives multilayering in particular, may offer further quality potential for SCO and SCR grades, but is still in the testing phase.

Supercalendering is one of the core parts of the manufacturing process. SC papers are traditionally calendered with 10–12 roll supercalenders, two or three per paper machine.

SC-B

SC-B grades are used for less demanding end uses than SC-A grades such as in low-end magazines (TV magazines and listings, for example), newspaper supplements, low-end catalogs, and also in freesheets in North America. SC-B grades are printed in rotogravure and in offset. Its competitors are special newsprint and also standard SC.

Basis weights range from 40 to 70 g/m^2, with the most common being 52–60 g/m^2. SC-B differs from SC-A grades mainly in brightness and in surface smoothness; it is less bright and rougher. For 52 g/m^2 SC-B, ISO-brightness is on the level of 67%–69%. The quality varies a lot between different producers, depending on the raw materials used as well as on the product equipment. The quality level has been approaching that of standard SC especially in Europe.

SC-B grades have slightly higher mechanical pulp content, from 75% to 90%, and lower chemical pulp content, from 10% to 25%, than the SC-A grade. Part of the mechanical pulp is often replaced by recycled pulp – up to 25%–30%. Filler content is clearly lower, 8%–15%, than in SC-A grades.

SC-B paper is produced either by old SC machines or by modernized newsprint machines. Finishing is achieved either through supercalenders or by a soft-nip calender. Soft-nip calendering is gaining more ground.

SC-C

SC-C paper is a North American uncoated paper grade, close to SC-B, where mechanical pulp dominates. Regarding properties and end use applications, it lies somewhere between newsprint and SC-B papers. It is often manufactured by modernized newsprint machines. A soft-nip calender is often used for the finishing. Its overall supply is sensitive to the market situation. This paper grade competes with special newsprint grades and also with SC-B grades.

2.1.3 Coated mechanical papers

Coated mechanical papers form a wide and increasing group of printing papers, where the base paper consists mainly of high-quality mechanical pulp and where long fiber chemical pulp is used to give it strength. The development of coating technology, new coaters, and new raw material mixtures will probably increase the number of these grades used in the future. The majority of coated mechanical papers are manufactured and sold in reels. Higher basis weight grades such as MWC and HWC are also available in sheets. Figure 17 illustrates relative positions of coated, mechanical pulp dominating papers in basis weight and brightness coordinates.

Printing and writing papers

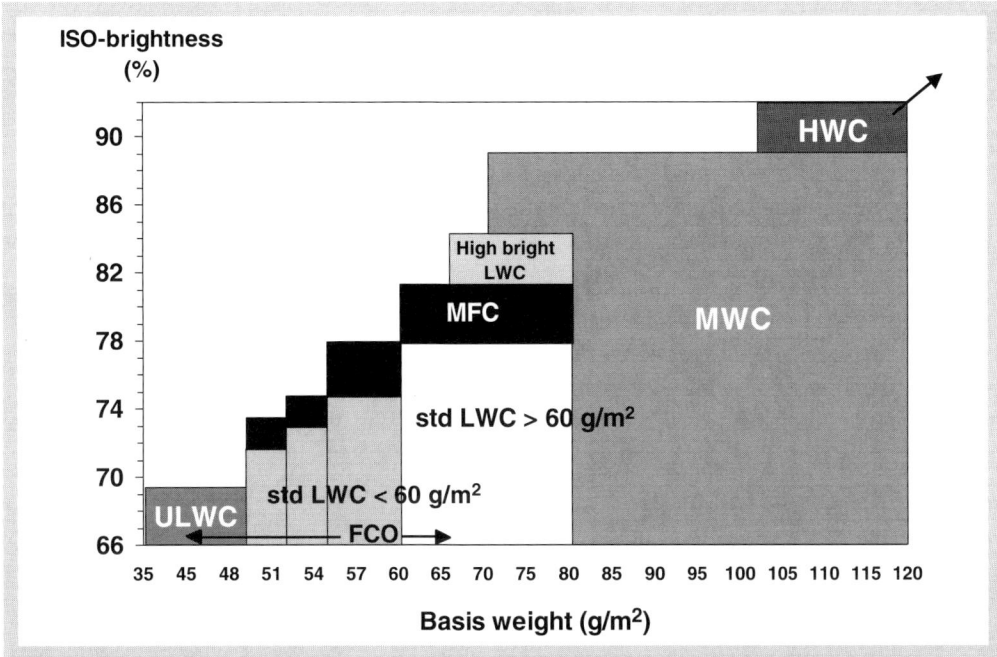

Figure 17. Positioning of coated, mechanical pulp dominating paper grades according to basis weight and brightness.

Standard LWC papers

LWC paper is a light weight coated paper, where the coat weight varies from 5 to 12 g/m²/side. Typical end uses for LWC papers are magazines, catalogs, inserts, and commercial printing. Figure 18 shows typical end uses for LWC and MFC papers in Western Europe. Figure 19 presents typical end uses for coated # 4 and # 5 in the United States. Coated # 5 in the United States is the closest corresponding grade to the standard European LWC paper; whereas, in # 4

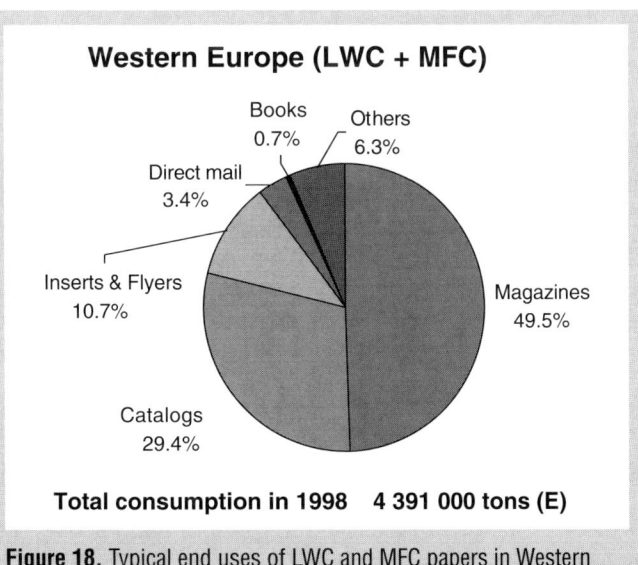

Figure 18. Typical end uses of LWC and MFC papers in Western Europe in 1998[8].

CHAPTER 1

grade, mechanical fiber content may vary from slightly mechanical to the level of standard LWC. Coated # 4 is brighter than coated # 5 as described later in the Section "American paper grade classification."

LWC papers compete with other mechanical pulp dominating papers such as FCO papers, SC-A+ papers (e.g., catalog end use), but also with coated fine papers in commercial printing applications. LWC grades are produced for both heatset web offset and gravure printing. Although offset printing is growing faster than rotogravure (as in magazines, for example, due to a more targeted audience), the advantage of gravure lies in its ability to produce significantly more pages per hour and cylinder set than web offset. Production speed in rotogravure is about double compared to offset, 20 m/s vs. 10 m/s. In rotogravure, a 144-page magazine can be printed in one run, whereas it requires four separate runs in offset to complete the job.

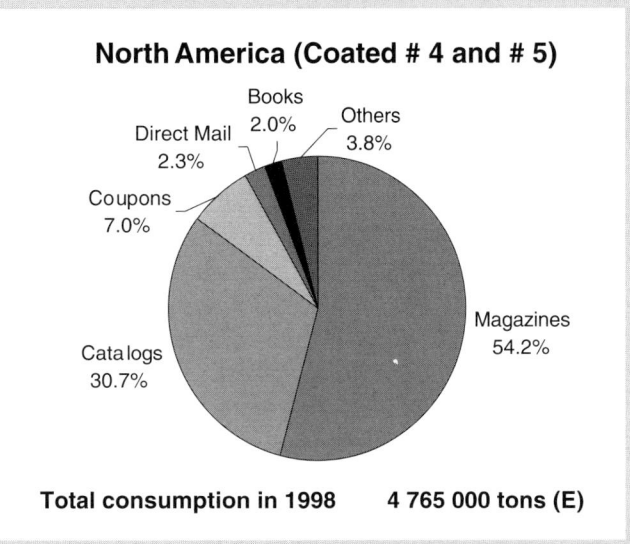

Figure 19. Typical end uses for coated # 4 and # 5 in North America in 1998[9].

Basis weight varies for LWC offset (LWCO) from 39 to 80 g/m² and for rotogravure (LWCR) from 35 up to 80 g/m². In 60 g/m² LWCO paper, ISO brightness is >72% up to 76%. Other important properties are good smoothness with minimal roughening tendency when drying printing ink as well as print gloss. Adequate stiffness is also important for LWCO. Excellent smoothness and compressibility are important for LWCR as well as excellent profiles due to the use of large reels. The biggest rotogravure reels today are about 3.6 m wide, are 1.35 m in diameter, and can weigh up to 7 t. Handling, transporting, and running such a large reel presents a real challenge for the whole chain. Heavy LWCR grades, in particular, are available also with matte finish.

Fiber furnish comprises 50%–70% mechanical pulp, groundwood pulp (GW or PGW), or thermomechanical pulp (TMP or CTMP), and 30%–50% chemical pulp. Base paper contains 4%–10% pigments. The total pigment content varies from 24% to 36%. Typical filler pigments are clay, talc, and also calcium carbonate ($CaCO_3$) in the case of a neutral process. The most common coating pigments are clay or clay and $CaCO_3$ for offset grade and clay or clay and talc for rotogravure grades. Special pigments are also used as additional minerals to give extra opacity and brightness as well as plastic pigments to improve gloss.

Printing and writing papers

In modern LWC base paper machines the web is formed by hybrid or gap formers. The latest innovations in wet pressing are shoe presses, which are applied for coated and uncoated mechanical grades. By using this new technique, the dry content of the web can be increased and thus save energy. Some improvements in bulk and stiffness have been reported.

The most frequently used coating technology for LWC papers is blade coating. One of the two main techniques in use in blade coating is short dwell time applicators (SDTA) – these were introduced in the early the 1980s, with the latest modifications being from the mid-1990s. This technique is suitable for low coat weights. Valmet has developed an Optiblade-concept and Beloit has developed Exel. Another blade coating technique is a long dwell time applicator (LDTA) used especially for higher coat weights. A new generation of free jet coaters was introduced in the mid-1990s for high-speed machines. Voith has developed Jet-Flo and Valmet Opticoat Jet. Film coating has recently established its position as a coating method in standard LWCO production.

Figure 20 shows some coating concepts for coated printing and writing papers. On-machine coating is a common way of producing LWC papers today.

LWC base paper is normally precalendered before coating to stabilize thickness and to reduce roughness and porosity. Final calendering is traditionally done with 10–12 roll supercalenders in Europe and with 9 roll supercalenders in the United States.

Coating concept	Coated paper grade	Basis weight g/m²
Film coater	ULWC, LWC, FCO	< 60
SDTA	LWC	(35) 40–60
LDTA	LWC	(35) 50–70
Film + LDTA	MWC, WFC	70–100
SDTA + LDTA	MWC, WFC	75–120
LDTA + LDTA	MWC, WFC	80–150
Film + LDTA + LDTA	WFC	100–170

Figure 20. Coating concepts for different paper grades[7].

Matte LWC papers can be finished by soft calenders or by-passing some nips in the supercalender. First on-line calenders have been ordered for manufacturing glossy LWC. See Ref. 10 for additional information on coated papers.

ULWC papers

Ultralight weight coated papers (ULWC), also sometimes known as low lightweight papers, are used in Europe for rotogravure printed catalogs and in the United States in offset printed magazines. Basis weight area ranges normally from 35 to 48 g/m². Adequate opacity is important especially in the United States because of strike-through. ISO brightness is at the level of 69%.

CHAPTER 1

MWC papers

Medium weight coated (MWC) papers are mechanical pulp dominating papers, where the coat weight amounts to 12–25 g/m^2/side. MWC papers are also known as double coated papers, whose information carrying capacity is higher than that of LWC paper. MWC papers can also be coated only once. A more homogenous surface structure leads itself to good smoothness and creates good printed ink gloss. They are used for high-quality special interest magazines with high-power advertisements, but also for catalogs, direct mail, and for other advertising. Figure 21 presents typical end uses for MWC papers in Western Europe. MWC papers compete with coated fine papers. They are mainly produced as an offset grade.

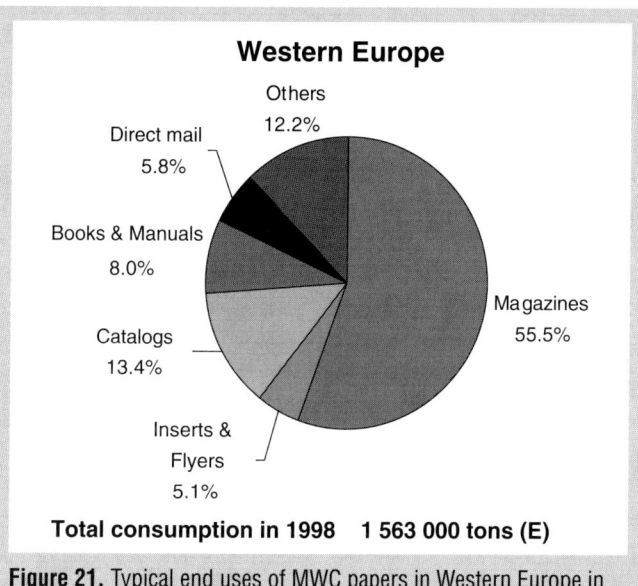

Figure 21. Typical end uses of MWC papers in Western Europe in 1998[8].

The basis weight varies between 70 and 130 g/m^2, with 70–90 g/m^2 being the most common weight. Important properties are high brightness, good opacity and gloss, excellent smoothness, adequate enough stiffness, and no cracking when folding. The coated surface of this grade provides minimal ink absorbency, which reduces the dot gain achieved in offset printing and enhances the definition of the print result. MWC papers are produced both in matte and gloss finish.

Base paper is made from 40%–55% mechanical pulp and 45%–60% chemical pulp. The total pigment content varies from 28% to 45%.

MWC papers are often produced with lower speeds than standard LWC.

HWC papers

High weight coated (HWC) papers differ from other coated, mechanical pulp dominating paper grades in their higher coat weight. Traditionally they are produced at the basis weight level of 100–135 g/m^2, and they compete with coated fine papers at such end use areas as high-quality magazines, catalogs, and magazine covers as well as in the area of direct advertising. HWC papers can be double or triple coated.

MFC papers

Machine finished coated (MFC) papers have higher bulk, are stiffer than LWC papers, and usually have a matte surface. The coat weight is low and varies from 2 to

10 g/m²/side. MFC papers are used for special interest magazines, catalogs, commercial printing, and books. MFC is a good option in situations where high readability, high delta gloss (printed gloss vs. paper gloss), a good opacity and brightness combination, and stiffness are needed. They are printed in heat-set web offset. MFC papers compete with LWCO, SCO, and machine finished specialties.

Basis weight varies from 48 to 80 g/m². Important properties of MFC papers are good surface properties, high print gloss, and adequate sheet stiffness.

MFC papers are made up of 60%–85% groundwood or TMP and 15%–40% chemical pulp. The total pigment content is 20%–30%.

MFC papers are often manufactured on an old, but modernized machine with an on-line blade coater or a film coating unit, but also a size press or gate-roll coater have been used. Soft-nip calendering is gaining more ground, but supercalenders can also be used. See Ref. 11 for additional information on MFC papers.

FCO papers

Film coated offset (FCO) paper is a fairly new paper grade. Film coating technology was introduced as a coating method for mechanical pulp dominating papers in mid-1990s in Europe. Film coaters have also been ordered to some of the latest LWC-lines. FCO is used, for example, in special interest magazines and catalogs. As the name suggests, it is good for heat-set web offset printing and competes mainly with LWCO grades.

It is currently produced in the basis weight range of 45 to 65 g/m². This grade gives a high bulk, good surface strength with low fiber roughening, and sufficient gloss compared to LWCO; however, coating leaves the surface rougher than in blade-coated paper due to a different surface structure. FCO is today very close to LWCO in regard to paper quality. Reference 11 provides additional information on FCO papers.

2.2 Chemical pulp dominating paper grades

Chemical pulp dominating paper grades comprise uncoated or coated fine paper grades, which generally contain only traces of or no mechanical pulp and 5%–25% fillers. According to the U.S. customs tariff, maximum mechanical fiber content is 10%. Both hardwood and softwood chemical pulps are used. The best properties of these papers are good strength, high brightness, and good archival characteristics.

Recently launched high-quality fine paper grades contain relatively high amounts of mechanical pulp made of aspen or poplar.

Chemical pulp dominating paper grades are usually produced in a neutral or alkaline papermaking process giving good strength properties to the web. Neutral papermaking conditions allow to use $CaCO_3$ as a pigment and produce good archival properties to the paper. High amounts of fillers and hardwood pulp give good formation to the paper.

New key technologies in use are shoe presses to increase machine speed and the dry content of the paper before the drying section, and gap former technology for better formation and drainage. Additive multilayering technology can also offer quality potential in these grades.

Uncoated fine papers are surface sized or pigmented, today normally with sizers. Coated fine papers are either pigmented (low coat weight papers) or first pre-coated with a sizer and thereafter coated (standard coated fine paper). These papers can even be triple coated (art paper).

Traditionally uncoated fine paper grades are calendered in paper machines with an on-line machine calender operating with one or two nips. Today's technology is moving toward soft calendering. In regarding coated fine paper grades, precalendering more often these days than soft calendering is gaining ground. Supercalendering follows precalendering. Matte finished grades are typically calendered on-line in conjunction with a coating machine using one or two soft-soft nips.

Fine papers are available in reels, sheets, or in both depending on the end use. Typically paper grades used for copying purposes are available in A4-size sheets or in bigger sized sheets for posters. Paper used for commercial printing is available mainly in reels, as is paper used for forms and listing paper.

Fine paper customers differ from those of mechanical pulp dominating papers in some key respects: Merchants play an important role in the fine paper business, taking care of storing the paper and service of final customers. Secondly, office equipment manufacturers (OEMs) form an important customer segment, which offers a unique opportunity for new product and quality development. Normally office equipment suppliers test the paper grades and give the approval for a paper to be used in their machines. OEMs monitor quality level continuously.

2.2.1 Uncoated fine papers

Offset papers and lightweight papers come under the classification of uncoated fine papers and are also known as woodfree uncoated papers (WFU papers). Soft calendering is gaining in popularity when producing these grades. Figure 22 illustrates typical end uses for uncoated woodfree papers in Western Europe, and Fig. 23 shows to same for uncoated woodfree papers in North America (also called uncoated freesheet). Writing papers belong to this group. They are sized and often slightly calendered. Also a filler content is low. A surface strength of writing papers is one of the most important properties.

Western Europe

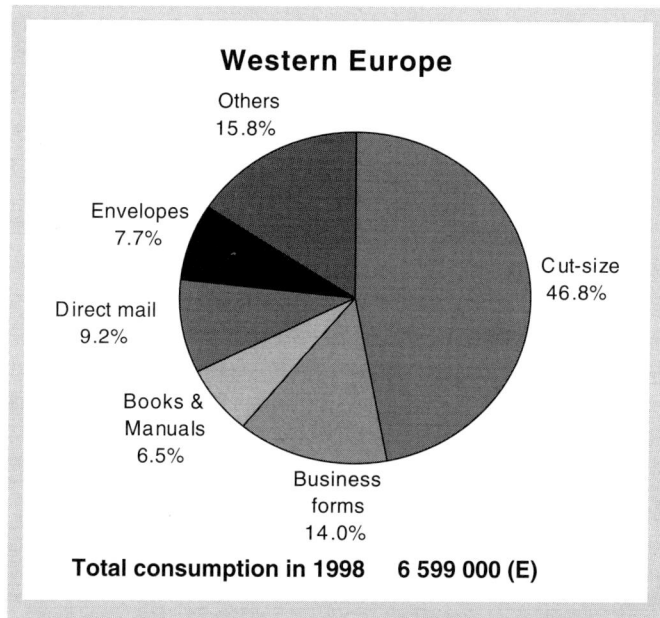

- Others 15.8%
- Envelopes 7.7%
- Direct mail 9.2%
- Books & Manuals 6.5%
- Business forms 14.0%
- Cut-size 46.8%

Total consumption in 1998   6 599 000 (E)

Figure 22. Typical end uses of uncoated WF papers (excluding carbonless papers) in Western Europe in 1998[8]

Printing and writing papers

Offset papers

Offset grades consist of a large variety of machine finished papers with various end uses. For example, 70–90 g/m² offset papers are used for commercial printing, books, magazines, and catalogs. Although 85% of these papers are sold in sheets, they are also available in reels.

Fiber furnish consists of more than 90% chemical pulp. Total pigment content varies from 0% to 25%.

The basis weight range is wide and it varies from 40 to 300 g/m². ISO brightness is > 80%. Surface smoothness is adjusted according to the end use. Further development of offset papers has mainly improved brightness and opacity.

Surface strength and low linting tendency are important properties for offset papers. The surface of offset papers is sized with starch, normally 0.5–2 g/m²/side. Additive multilayering technology has given some promising results for higher filler content offset papers, too.

Lightweight papers

Lightweight papers are lowweight offset papers. Basis weight varies from 25 to 40 g/m². They are used mainly for direct marketing and can be either surface sized or pigmented. Among the lightweight papers, bible and dictionary paper form the largest product group.

North America

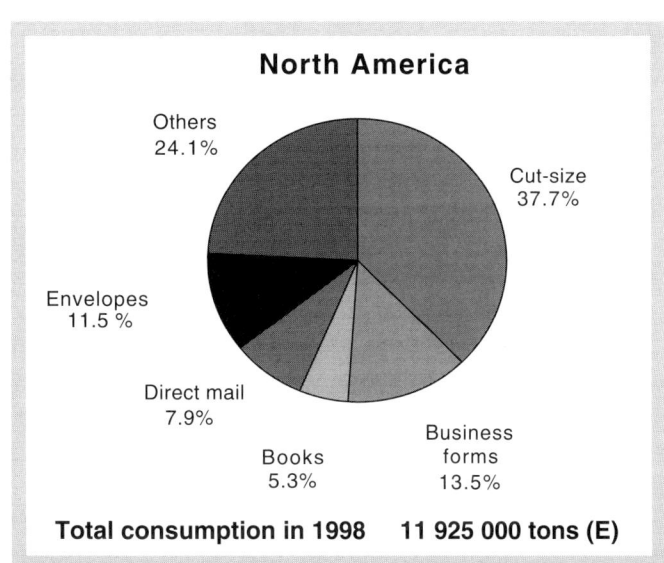

- Others 24.1%
- Cut-size 37.7%
- Envelopes 11.5 %
- Direct mail 7.9%
- Books 5.3%
- Business forms 13.5%

Total consumption in 1998   11 925 000 tons (E)

Figure 23. Typical end uses of uncoated WF papers in North America in 1998[9].

Western Europe

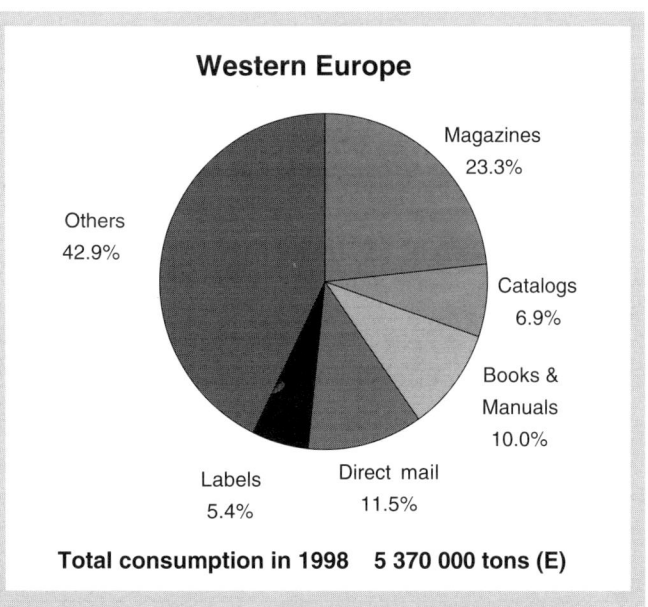

- Others 42.9%
- Magazines 23.3%
- Catalogs 6.9%
- Books & Manuals 10.0%
- Direct mail 11.5%
- Labels 5.4%

Total consumption in 1998   5 370 000 tons (E)

Figure 24. Typical end uses of WFC papers in Western Europe in 1998[8].

2.2.2 Coated fine papers

Coated fine papers consist of standard coated fine papers, WFC papers, low coat weight papers, and art papers. These papers are used for numerous demanding printing applications. Figure 24 presents typical end uses for coated woodfree papers in Western Europe, and Fig. 25 illustrates typical end uses for coated woodfree papers in North America (also called coated freesheet). They can be single, double, or triple coated. The surface can be machine finished and matte or gloss calendered. Jet coating technology is rapidly gaining ground in coated fine papers similarly to how it is in coated mechanical pulp dominating papers.

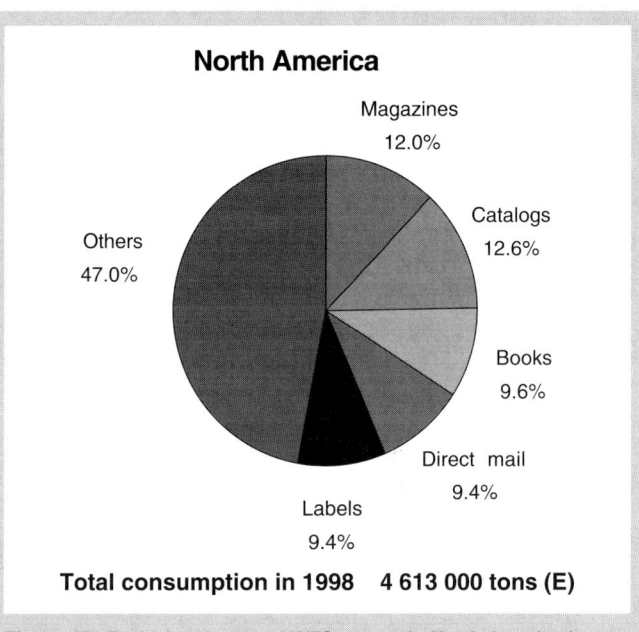

Figure 25. Typical end uses of WFC papers in North America in 1998[9].

Standard coated fine papers

Standard coated fine papers are used for advertising materials, books, annual reports, and high-quality catalogs as well as for magazines and, increasingly, direct mail. Competition comes from high bright LWC papers and from MWC papers in less demanding end uses and from art papers in the most demanding end uses. Standard coated fine papers are produced mainly for offset printing. Although these papers are produced for the most part (85%–90%) in sheets, they are also produced in reels.

Basis weight area ranges from 90 to 170 g/m^2 and ISO brightness varies between 80% and 96%. Fiber furnish consists of more than 90% chemical pulp. Total pigment content varies from 30% to 45%. Calcium carbonates and clay are the most commonly used pigments.

WFC fine papers are coated in different manners: for example, for precoating a sizer is used, which is followed in turn by a two-head on-machine coater. The most common way is to use an on-machine sizer followed by a double coating with a four-head off-machine coater.

Printing and writing papers

Low coat weight papers

Low coat weight papers are used for less demanding applications than standard coated fine papers such as books, directories, timetables, and brochures. Basis weights vary between 55 and 135 g/m². Coat weights are at a level of 3–14 g/m²/side. Total pigment content varies from 20% to 35%. Low coat weight papers are available both in sheets (70%) and also in reels. Single coated grades are produced with on- or off-blade coaters and pigmented grades with sizers.

Art papers

Art papers represent *one of the highest quality printing papers* and are used for illustrated books, calendars, and brochures. The basis weight varies from 100 to 230 g/m². Art papers are produced with matte or glossy finish. The coat weight varies from 20 to > 40 g/m²/side. Art papers are almost exclusively available in sheets. They are triple coated, first with a precoater and thereafter with a double blade coater. A new feature is a manufacture of art papers by new, effective, and big paper machines.

2.2.3 Special fine papers

Special fine papers include copy papers, digital printing papers, and continuous stationery.

Copy papers

Copy papers are used for copying and non-impact printing. Basis weight varies between 70 and 90 g/m² and ISO brightness between 80%–96%. The most important properties for a copy paper are smooth run in a copying machine and good dimensional stability. It must not show curling or cockling and retain dust when copying. It is mainly made of 90%–100% virgin chemical pulp fibers, but it may contain recycled fibers up to 100%. The recycled fibers used are mainly copy paper waste. Total pigment content varies between 10% and 25%.

Digital printing papers

Digital printing papers (synonymous with electronic printing papers) are a rapidly growing group of chemical pulp dominating papers that are mainly uncoated but increasingly coated fine paper grades. Opportunities for increasing digital printing are based on on-demand, short run printing and variable imaging as well as database printing and distributed printing[12]. Benefits stem from reduced lead times, minimal make-ready times and the ability to make every print different. Digital printing is expected to gain about 20% market share by the year 2010 largely at the expense of litho. The market share of digital printing is at the level of 7%–8 %[13]. Production speed is one of the current bottlenecks of digital printing systems due to insufficient computing power, but the problem will be solved most likely in the near future. Expanded use of digital printing will have an impact on the future development of whole printing industry and on the entire value chain. The consumption of digital printing papers was about 90 000 t in 1998, 75% of which was coated and 25% uncoated fine papers. This figure is expected to reach 250 000 t in 2003, of which more than 50% is expected to be coated. The bulk of this paper is delivered in sheets.

Digital printing technology is not a single technology, but almost every manufacturer has a different process. The major categories of nonimpact printing are electrical charge based methods, magnetic methods, thermal methods and ink-jet methods, which are described in details in another volume of this book series, Printing.

Requirements for printing papers vary by printing method. In electrophotography, the runnability of the paper is critical. The electrical conductivity of the paper should be sufficient as well as electrical resistivity. Moisture content and friction are also important properties. Other characteristics of importance are good dimensional stability to avoid curl and cockle for example, surface strength and surface smoothness especially with high resolution. Chemical properties of paper surface such as surface energy for example are important for fixing the toners to the paper, not for print quality as such. According to the study of Kulmala et al.[14] microscopical roughness and porosity, grammage and density were found to be of importance. In ink-jet printing, papers require characteristics that are matched with the inks and the drop volumes. First of all, ink-jet papers must be smooth. They must have sufficient and even porosity, composed of small pores, in order to absorb the solvent quickly and to counteract the spreading tendency. Also dimensional stability is important as well as cockling and curling tendencies. With a widenig printing paper range, digital printers have reported the following problems with paper: static electricity with coated papers and films, curl and moisture problems, and cutting problems in the heavier weights such as 190 g/m^2 [15].

Paper costs play a minor role in digital printing. Product costs consist mainly of inks, special films, and of printers themselves and typically do not decrease when the number of printed copies increase. Problems in recycling and decolorization of digital papers have not been solved to date.

Typically printing machine suppliers certify the papers accepted for their machines. Typical uses for digitally printed products are manuals, price lists, and various direct mail materials, but also low-volume paperbacks and hard-cover books. Black-and-white applications cover more than 80% of digital paper need at present. Basis weights vary a lot, from 40 to as much as 400 g/m^2.

Continuous stationery

Continuous stationery papers are used for listings and custom-made forms. Basis weights vary between 40 and 90 g/m^2. Continuous stationery must have good strength, purity, and dimensional stability. Pigment content varies from 5% to 25%. These papers can also contain some mechanical pulp. Continuous stationery is usually printed with on-line impact printers but also with non-impact laser printers.

2.3 Printing and writing paper grades: general trends

Table 3 describes general development trends for printing and writing papers. Many of them are customer-driven such as, for example, a demand for higher brightness in Europe, but some trends are also final end consumer-driven such as the increasing use of recycled fiber.

Table 3. General trends of printing and writing papers.

```
new end uses
new paper grades
        → difficulty to classify paper grades
changing paper characteristics
        higher brightness
        lower basis weights
        polarization of properties
changes in raw materials
        increasing use of RCF
        expanding use of minerals
```

Advertisers try to improve the effect of an advertisement through an increased contrast between the printed and unprinted areas resulting in *higher brightness* demands.

Pressure to cut distribution costs, to use fewer resources and to increase the printing surface drive for *lower basis weights*.

Polarization of properties such as basis weight (high weights vs. ultralight) and gloss (gloss vs. matte) seems to be yet another trend. The growth in number of paper grades, however, satisfies increasing and diversifying customer needs.

The emergence of new paper grades is attributable to many different reasons such as:

- Customers' diversifying needs boosted by increased target marketing
- Advancing printing technology such as digital printing creating *new end uses*
- Advances in paper manufacturing technology, especially in the areas of multi-layering, coating, and finishing
- Producers' upgrading needs of older machinery due to profit erosion also speed up this development
- Increasing usage of recycled fiber, especially in newsprint and SC-B grades
- Increasing usage of different minerals and mineral combinations.

The emerging new paper grades have created *a paper classification problem*. There are, for example, a few slightly chemical pulp dominating papers, which contain significant amounts of mechanical fibers and are difficult to find a niche for instance in the European classification system.

CHAPTER 1

Another trend is *the increasing use of recycled fiber as a raw material* in newsprint and in SC-B grades in particular. Availability and price of recycled fiber has an impact on this trend.

Along with the consolidation development of the paper industry and growing diversity of paper machines within a company, *new efficient paper machines tend to concentrate on a narrow product range* to both maintain an even quality level and to guarantee high production efficiency.

Table 4 illustrates the typical mineral contents of the end uses of printing and writing papers.

Table 4. European printing and writing papers: mineral content and end uses[16].

Mineral content	Production capacity in 1996 million metric t	End uses
3%–15%	Newsprint 10.3	Newspaper, inserts, flyers, telephone directories
5%–35%	Unctd. mech. 6.3	Magazines, supplements, catalogs, inserts, books
28%–48%	Ctd. mech. 8.3	Magazines, catalogs, brochures, direct mail
25%–50%	Ctd. WF 8.6	Magazines, brochures, direct mail, annual reports, books
15%–30%	Unctd. WF 10.6	Office papers, writing papers, envelopes, books

3 Different regional paper grade classifications

3.1 European paper grade classification

The paper grade classification of printing and writing papers detailed throughout this chapter follows the European system. The key classification criterion is the pulping method of the main fiber component. In this presentation, this means the distribution of mechanical pulp dominating paper grades and chemical pulp dominating paper grades. Alternative terms in use are "mechanical printing papers" and "woodfree printing and writing papers" for the same purpose. The present European classification is very production-driven and reflects more precisely, for example, the coat weight such as ULWC, LWC, MWC, and HWC papers. This is also true of coating methods as in the case of FCO and finishing technology such as in SC, MFC, and MFS papers. It can also happen that when a new paper grade is first given a name, for instance, the finishing method used at the time, it can later be manufactured according to a different method, but still reach the same quality standards. It is seldom that the actual grade name indicates its ultimate end use such as a telephone directory (TD paper), for example. Additional determinants such as R in the case of SC<u>R</u> and O in the case of LWC<u>O</u> are used to indicate the printing method. (R indicates rotogravure, and O indicates offset.) SC <u>Cat</u> describes the end use application – a sales catalog. The main problem with the current classification system is that it does not show how well a product meets a customer's specific need. In practice, this is overcome through a close cooperation between suppliers and publishers and printers as well as merchants.

3.2 American paper grade classification

In the United States, coated printing papers are defined in a different manner. This classification is based upon the brightness of a paper and is numerical. Coated papers are classified according to the American Forest Products Association (AFPA)[17]:

Coated paper grade	Brightness (GE)
Premium	88 and above
Number 1	85.0 to 87.9 (inclusive)
Number 2	83.0 to 84.9 (inclusive)
Number 3	79.0 to 82.9 (inclusive)
Number 4	73.0 to 78.9 (inclusive)
Number 5	72.9 and less

Between Premium and Number 2 above, there are chemical pulp dominating grades. From Number 3 onward the amount of mechanical pulp increases. Dull, matte, and glossy grades are all classified according to brightness. The classification does not make any distinction between different finishing methods.

Coated printing paper meets the following criteria:

- Papers that have a surface coating to improve the appearance and printing surface
- Papers less than 50 lb in basis weight, with coat weights per ream (sheet size: width 25 in. x height 38 in., 500 sheets) of not less than 2.5 lb per side with 50% or more of the added coat weight consisting of pigment
- Papers 50 lb or heavier in basis weight with coat weights per ream (25 in. x 38 in. -500) of not less than 4 lb per side with 50% or more or of the coat weight added consisting of pigment
- Papers that are manufactured in basis weights up to 110 lb or to 120 lb if the weight(s) higher than 110 lb represents a continuation of the range of basis weights in which the grade is manufactured (25 in. x 38 in. -500).

3.3 Japanese paper grade classification

The Japanese printing paper classification follows another logic. It is based on dominating pulp grade (groups A and B) and the amount of coating (1, 2, and 3 after a letter) "A" means chemical pulp dominating grades, and "B" mechanical pulp dominating paper grades[18].

CHAPTER 1

Table 5. Japanese paper grade classification.

Coating	Japanese classification	European classification
Uncoated	Printing A	WF uncoated papers, 100% bleached chemical pulp,
	Printing B	Semimechanical printing papers, >70% bleached chemical pulp, brightness about 70%
	Printing C	Mechanical printing papers, 40%–70% bleached chemical pulp, brightness about 65%
	Printing D	Mechanical printing papers (standard), less than 40% bleached chemical pulp, brightness above 55%
	Printing E	Newsprint based on deinked pulp, even 100% DIP, used for comic magazines, often tinted
	Printing G	Supercalendered gravure paper, containing mechanical pulp
Coated	Coated A1	Art coated papers (triple coated), 84.9–157 g/m^2, mainly for sheetfed offset, 20 g/m^2 coat weight/side
	Coated A2	WF coated (double coated), 73.3–157 g/m^2, mainly for sheetfed offset, 10 g/m^2 coat weight/side (also called coated papers)
	Coated A3	WF light coated (single coated), 58.7–157 g/m^2 (the main substance 64.0–81.4 g/m^2), mainly for heat set web offset, use in posters, catalogs and magazines, coat weight 7.5 g/m^2/side (WF "LWC")
	Coated B2	Mechanical coated papers (standard and improved), 60.2–81.4 g/m^2 (the main area up to 72.3 g/m^2), for both offset and rotogravure printed magazines, coat weight 10 g/m^2/side
	Coated B3	Mechanical coated papers (below standard), 60.2–81.4 g/m^2 the main area being 60.2–72.3 g/m^2, for offset and rotogravure, 5 g/m^2 coat weight per side
Slightly coated	Bitoko WF	Light WF coated, base paper: Printing A; brightness not less than 79%, supercalendered or machine finished, used for catalogs and magazines, coat weight below 6 g/m^2/side
	Bitoko 1	Light semimechanical, base paper: Printing B; brightness 74%–78% supercalendered or machine finished, used for catalogs and magazines, coat weight below 6 g/m^2/side
	Bitoko 2	Light LWC, base paper: Printing B; brightness 68%–73%, supercalendered or machine finished, use in magazines and catalogs, coat weight below 6 g/m^2/side, used for catalogs and magazines
	Bitoko 3	Light surface treated mechanical paper, base paper: Printing C; brightness 62%–67%, supercalendered or machine finished, coat weight below 6 g/m^2/side, use in catalogs and magazines

Notes:

1. Uncoated paper grades from Printing A to Printing E all contain RCF to a varying degree, Printing E uses 100% RCF. In addition, real newsprint, which is sold to newspaper publishers, is not included in this classification but sold as a tailor-made, often surface-sized, product to a publisher.

Printing and writing papers

2. Uncoated Printing G is locally produced only to the limited extent.
3. Coated A2, known also as "full WF coated paper," is the main WFC grade for printing and writing.
4. Coated A3 is the main HSWO grade for commercial printing.
5. Coated B2 also known as "premium LWC."
6. Bitoko WF is for the main part high-bright matte paper for publishing, better quality than in traditional MFC paper.
7. Bitoko 1 and bitoko 2 are known as "Japanese LWC papers."
8. Bitoko 3 comparable to standard SC offset paper regarding printed quality.

3.4 Printing and writing paper grade classification according to FAO

FAO classifications follow those contained in *Classification and definitions of forest products, FAO, Rome, 1982*. Two main categories are *newsprint* and *other printing and writing paper*. FAO classification for printing and writing papers is as follows[19]:

- Newsprint
- Other printing and writing paper
    - Coated printing and writing paper
        - Coated wood containing printing and writing paper
        - Coated woodfree printing and writing paper
    - Uncoated printing and writing paper
        - Uncoated wood containing printing and writing paper
        - Uncoated woodfree printing and writing paper.

The group "newsprint" is defined more precisely in the following manner:

641.1 NEWSPRINT: "Uncoated paper, unsized (or only slightly sized) containing at least 60% mechanical wood pulp (% of fibrous content), usually weighing not less than 40 g/m^2 and generally not more than 60 g/m^2, used mainly for printing newspapers".

The other main group "printing and writing paper" is defined as follows:

641.2/3 PRINTING AND WRITING PAPER: "*Other printing and writing paper.* Paper, except newsprint, suitable for printing and business purposes, writing, sketching, drawing, etc., made from a variety of pulp blends and with various finishes. Included in this group are such papers as those used for books and magazines, wallpaper base stock, box lining and covering, calculator paper, rotonews, duplicating, tablet or block, label, lithograph, banknote, tabulating card stock, bible or imitation bible, stationery, manifold, onionskin, typewriter, poster etc."

FAO's printing and writing paper classification differs from other known classifications in the respect that it also includes some special paper grades that, for example, have been assigned to special papers in this book.

CHAPTER 1

3.5 Printing and writing paper classification according to CEPI

Confederation of European Paper Industry (CEPI) classifies printing and writing papers and refers to them as graphic papers as follows[20]:

Newsprint: Paper mainly used for printing newspapers. It is made largely from mechanical pulp and/or recovered paper, with or without a small amount of filler. Weights usually range from 40 g/m^2 to 52 g/m^2 but can be as high as 65 g/m^2. Newsprint is machine finished or slightly calendered, white or slightly colored, and is used in reels for letterpress, offset, or flexo printing.

Uncoated Mechanical: Paper suitable for printing or other graphic purposes where less than 90% of the fiber furnish consists of chemical pulp fibers. This grade is also known as groundwood or wood-containing paper and magazine paper, such as heavily filled supercalendered paper for consumer magazines printed by the rotogravure and offset methods.

Uncoated Woodfree: Paper suitable for printing or other graphic purposes, where at least 90% of the fiber furnish consists of chemical pulp fibers. Uncoated woodfree paper can be made from a variety of furnishes, with variable levels of mineral filler and a range of finishing processes such as sizing, calendering, machine glazing and watermarking. This grade includes most office papers, such as business forms, copier, computer, stationery, and book papers. Pigmented and size press "coated" papers (coating less than 5 g/m^2 per side) are covered by this heading.

Coated Papers: All paper suitable for printing or other graphic purposes and coated on one or both sides with minerals such as china clay (kaolin), calcium carbonate, etc. Coating may be by a variety of methods, both on-machine and off-machine, and may be supplemented by supercalendering.

3.6 Some comments on different classification systems

Different classifications cause problems and often confusion regarding the use of statistics. Regional classification systems are based on a local grade supply. Industry statistics are available such as those from official organizations, consultancy companies, industry analysts, and of course paper industry companies themselves following different classification criteria. Now that the number of paper grades is increasing, with classification being practiced on different grounds and with paper trade still globalizing, the situation is becoming even more difficult to monitor. Reference 10 provides additional information on different classification systems.

4 Some aspects on choosing a paper grade for a specific purpose

The increasingly diverse number of paper buyers are faced with a bewildering choice of suppliers and brands. Manufacturers are responding to changing requirements by bringing out a stream of new products. There are many factors to consider when choosing the most suitable paper grade for a specific purpose. Figure 26 presents key drivers for choosing the most suitable printing and writing paper grade for a specific end use from the publishers' and printers' viewpoints.

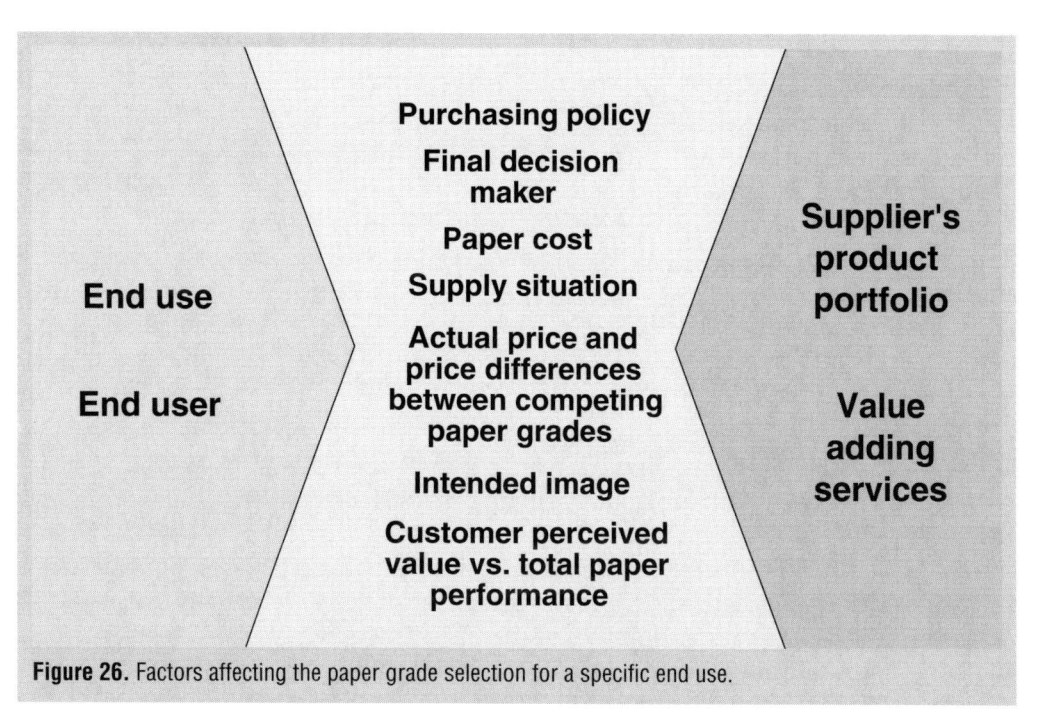

Figure 26. Factors affecting the paper grade selection for a specific end use.

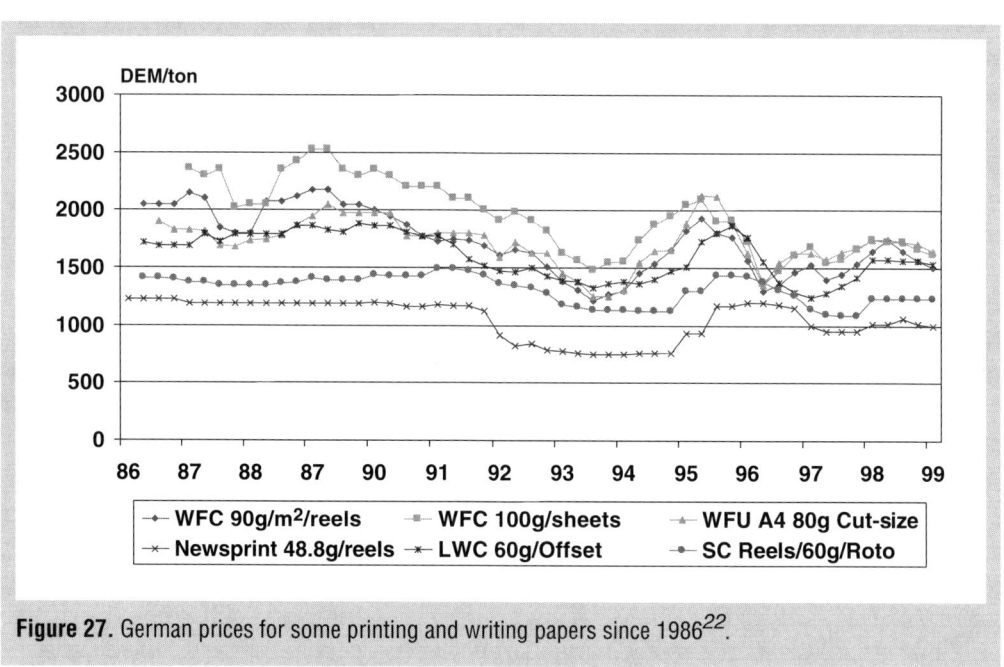

Diagram elements:
- End use
- End user
- Purchasing policy
- Final decision maker
- Paper cost
- Supply situation
- Actual price and price differences between competing paper grades
- Intended image
- Customer perceived value vs. total paper performance
- Supplier's product portfolio
- Value adding services

Figure 27. German prices for some printing and writing papers since 1986[22].

Legend:
- WFC 90g/m² /reels
- WFC 100g/sheets
- WFU A4 80g Cut-size
- Newsprint 48.8g/reels
- LWC 60g/Offset
- SC Reels/60g/Roto

CHAPTER 1

The end user, such as a magazine publisher or a job printer, and *the final end use,* such as a rotogravure printed sales catalog on cosmetics, set the basic requirements for a paper grade.

The paper as a cost factor is also an important selection criterion. In products for mass distribution and in large printing jobs, the price of the paper has great significance. For example, paper used for newspapers, can account for 50%–70% of the total production costs, whereas this can only be at a level of 3% in a sales catalog.

Actual price and *price differences between competing grades* at a time of decision making also matter. Prices and price variations differ significantly depending on the cycle as Fig. 27 indicates. Price is not, however, a sole factor in paper choice. According to PIRA[21], quality, service, delivery reliability, availability of a product, and the range of products are of importance.

A final decision maker such as an advertiser, editor-in-chief, or a purchasing manager of a publisher has a strong say when selecting a paper for a specific purpose.

Purchasing policy of a customer also has an impact on papers used for different end uses.

The size of a supplier's product portfolio also has an impact on paper selection. Multinational publishers and independent printers seem to be growing in size as the paper industry is simultaneously increasing its global presence. So large paper companies with wide product portfolios and transnational manufacturing can in the easiest manner satisfy the customers' needs.

Supply situation, too, has effects on paper selection in this very cyclical industry. During an upturn, a delivery time of a preferred paper grade may be too long to satisfy some unexpected needs and that is why another paper grade is chosen.

It is not only the paper which matters but also the existing *value adding services* when making a decision as to which paper is to be used. For example, a printer might want to have some technical assistance in choosing an optimal printing ink when using a new paper grade for a job.

Customer perceived value vs. total paper performance as in reproduction capacity, runnability, ink consumption, and waste as well as weight, brightness, and opacity balance have increasing importance on paper selection. Many publishers and printers are continuously and systematically monitoring paper performance and comparing suppliers.

A chosen paper grade must support and be in balance with the content of an end use and advertisements. So *too does intended image of the final product* have an impact on paper selection.

Paper can communicate rationality or economy, high technology, or ecological values. Different target groups make different evaluations about paper.

Selecting a paper for a specific purpose is essentially a compromise between different criteria.

When designing a new product, the selection of paper grade should be followed by graphic design, not the other way round. The choice of paper is already a creative solution and is important for the final message of the printed product. Paper selection is in part rational decision-making, but the selection is usually imprecise. When a graphic designer selects a paper, the important factors are printed product image, target group, mode of use and purpose, structure, printing method, screen density, special effects, pictures, amount of text, price, paper availability, and timetable. Paper as such is already a message.

The number of paper grades is increasing due to customers' diversifying needs and paper manufacturers' upgrading needs. Also new technological innovations and a competitive price of one of the key raw materials can boost an emergence of new paper grades. So there are both customer pull and technology push driving forces. The differences between different paper grades are diminishing. The operating mode "right paper for right purpose" requires a tight cooperation between the decision maker regarding paper and the supplier as well as continuous updating of knowledge due to rapid technological development on both sides. In addition to paper technical and printing properties, functional properties such as stiffness and bulk must be taken into consideration. Also the pleasantness of paper has a role to play. Reference 23 provides additional information on paper properties and pleasantness of paper.

5 Key drivers for substitution between printing and writing paper grades

In substitution between printing and writing paper grades are in effect two different types of forces: short-term and long-term ones. Both the short-term actual situation and the longer-term development affect the final outcome simultaneously.

The strongest short-term forces are the *supply/demand balance* and the *price differences between competing grades.* In an upturn when a supply situation is tight and price differences between grades are at their smallest, substitution might take place because a desired paper grade is not easily available. In a downturn, by contrast, when there is an excess of paper and the price differences are at their greatest, substitution might happen because of prominent cost savings.

Long-term influences for substitution are, for example, *decreasing quality differences* between printing and writing paper grades, *widening product range, development of printing technology* for a unifying effect, *development of paper production technology* with new coating and finishing applications, and *pressures from advertisers and consumers* as in the form of environmental demands[4].

6 Some comments on the challenge from electronic media

Print media has perhaps for the first time in its history a genuine long-term challenger, the electronic media. Increasing electronic information transfer and digitalization of all information are probably the core challengers of paper as a printing surface.

Rapidly developing *digital technology* creates a number of opportunities for both print and electronic media. Electronic media has benefited more from rapid technological development than print media. Continuously improving paper manufacturing technology, however, has created new paper grades and cheaper as well as better quality paper. The most important driver for the further development of print media and electronic media is *the behavior of the final consumers.* Will they appreciate more the creativity challenges

CHAPTER 1

and logical reasonings that reading for "learning" and "relaxation" offers than the opportunity to obtain new experiences through surfing the Internet? Reference 24 provides additional discussion on the future of paper in the information society.

Print media still has many strengths over electronic media for both the final consumer and the advertiser[25, 26]:

Print media's strengths are:	Electronic media's strengths are:
+ User access to choice	+ Real-time information
+ Low cost	+ Globally fast
+ Portability	+ Easy tailoring of needs
+ Familiar medium	+ Integration of several media possible
+ Visual environment	
+ Stimulating reading experience	+ Search, sort, paste capabilities
+ Environmentally friendly	+ Interactivity
+ Based on recyclable, renewable resource	
+ Easy as an advertisement media: It does not require interactivity	
+ Information content	
+ Slowness: comprehensive and satisfies	
+ It satisfies the "higher" needs of human beings.	

<u>Potential threats for print media</u> come from distribution of paper and printed products, changing consumer's habits in the medium/long term, environmental pressures, and also, to a certain extent, from the unclear future of print media. Publishers probably have to redefine the roles of print and electronic media. It is currently believed that print and electronic media will continue to integrate and, in this way, support each other's growth and value.

<u>The current bottlenecks facing electronic media</u> are: price of products for mass distribution, inadequate safety features, a lack of content-rich products, missing standards, and also the user-unfriendly way in the organization of www-addresses.

The challenge created by electronic media to print media varies according to the end use sector: traditional business forms, encyclopedias, and manuals have been hit first, whereas special interest magazines and commercial printing seem to have the safest future.

Printing and writing papers

DOCUMENT APPLICABILITY TO ELECTRONIC PRODUCTION

	CATALOGS	MAGAZINES	DIRECTORIES	BOOKS	FINANCIAL & LEGAL	MANUALS & TECH. DOC.
Most (Information/Search)	Business to business catalogs	Professional & scientific journals	Industrial directories association directories	Professional & scholarly books	SEC filings	High-tech
	Specialty catalogs	Newsletter	Government directories	Workbooks		
	Retail store catalogs	Regional edition magazines	Local directories	Textbooks	Prospective	Capital goods
	Showroom catalogs	Special edition magazines		Consumer books: trade	Legal docs	Consumer goods
Least (Entertainment/Browse)	Major catalogs	Mass market magazines	Telephone directories	Mass market paperbacks		

Appropriateness: Least (Entertainment/Browse) → Most (Information/Search)

Figure 28. Document applicability to electronic media.

Figure 28 illustrates document applicability to electronic media. For example, manuals and technical documents are easier to transform into electronic production than catalogs, within the catalog group business to business catalogs are more easily transformed than showroom catalogs.

As an example of the competition potential of new electronic products, the electronic book project of the Media Lab of MIT is described here[27]. The key components of this electronic book project are the electronic ink and the control system that pinpoints the signal to electronic ink capsule at a given position of the "digital page." In other words, the electronic ink and the guidance of control signals for the ink capsules actually form an inexpensive screen system. The ink is a dispersion of tiny particles inside microcapsules, and it can be changed from white to black simply by applying an electronic field to form a desired image. New information can be loaded when the old has been read. The initial applications might be, for example, street signs and electronic badges. In the longer term, it could serve in more demanding applications such as disposable TV screens[28]. Professor Joseph Jacobson from MIT has estimated that it will take some four or five years before this technology will be available to newspaper readers. See additional information on an electronic book in Refs. 29 and 30.

# References

1. Jaakko Pöyry Consulting, Helsinki.
2. Croon, I., Svensk Papperstid./Nordisk Cellulosa (7):10 (1998).
3. Kilpi, S., "Printing and Writing World Demand to the Year 2010," 1997 International Business Planning Conference Proceedings, Papercast, Paris.
4. Haarla, A., "end uses as Drivers for Paper Grade Development, Advancing Papermaking," 1997 Intertech Seminar Proceedings, Intertech, Frankfurt am Main.
5. World Graphic Papers, Consultancy Report, EMGE & Company, Great Britain, 1998.
6. Lindahl M.J., "Supplying Global Markets from Transnational Manufacturing," 1998 PRIMA Conference Proceedings, PRIMA, Lisbon.
7. Valmet Corporation, Helsinki.
8. European Gaptrack, Vol. 1, Consultancy Report, Jaakko Pöyry Consulting, Helsinki, 1998.
9. Gaptrack, Second Quarter Report, Consultancy Report, Jaakko Pöyry Consulting Inc., New York, 1998.
10. Boothby, C., PIMA's Papermaker **79**(5):38 (1997).
11. Klass, C.P., Pulp and Paper **72**(3):71 (1998).
12. Brett, G. and Birkenshaw, J., Short Run Digital Color Printing, PIRA International, UK, 1998.
13. Birkenshaw, J.W., Management and Technology 5(2):18(1999).
14. Paulapuro, H., Kulmala, A., and Oittinen, P., IS&T's Tenth International Congress on Advances in Non-Impact Printing Technology, 1994, pp. 446–470.
15. "Digital Color Printing in Europe: 1997–2003," Mikulski Hall Associates, UK, 1998.
16. Haarla, A., "Paper and Minerals: an Outlook for European Printing and Writing Paper Industry and its Markets," 1997 Euromin Conference Proceedings, Industrial Minerals Information Ltd, Barcelona.
17. American Forest Products Association (AFPA), Washington DC.
18. Japan Pulp and Paper Association (JPPA), Tokyo.
19. Food and Agriculture Organization of the United Nations (FAO), Rome.

20. Confederation of European Paper Industries (CEPI), Brussels.
21. Howkins, M., Paper Choice in the UK, Consultancy Report, PIRA International, UK, 1998.
22. Pricewatch, Pulp and Paper International (PPI), Brussels.
23. Mensonen, A., "Pleasantness of Paper," Master's Thesis, Helsinki University of Technology, Helsinki, 1996. (in Finnish)
24. Paulapuro, H., The Electronic Library **9**(3):135 (1991).
25. Hoppe, J. and Baumgarten, H.L., Wochenbl. Papierfabr. **125**(18):860 (1997).
26. Hoppe, J. and Baumgarten, H.L., Wochenbl. Papierfabr. **125**(19):918 (1997).
27. Comiskey, B., Albert, J.D., Yoshizawa, H., et al., Nature **394**(7):253 (1998).
28. Wisnieff, R., "Printing screens," Nature **394**(7):225 (1998).
29. Cline, C., "Would You Curl Up with an Electronic Book?," Seybold Report on Internet Publishing, 1998, p.32.
30. Kawai, H. and Kanae, N., "Microencapsulated Electrophoretic Rewritable Sheet (MC-EPS)," 1999 SID Conference.

In addition, Valmet Oyj, Voith GmbH, and Beloit Corporation provided the writer with various background materials. The author worked when writing the chapter for UPM-Kymmene Corporation, Helsinki, as a Director in the Business Development Department of Paper Divisions.

The author wishes to thank Professor Fumihiko Onabe, Laboratory of Paper Science, The University of Tokyo, for valuable comments on Japanese paper grade classification.

CHAPTER 2

Paperboard grades

1	**Cartonboards**	**56**
1.1	Boxmaking process	56
	1.1.1 Mechanical requirements	57
	1.1.2 Purity and cleanliness	57
	1.1.3 Printing properties	57
	1.1.4 Summary of boxboard requirements	58
1.2	Folding boxboard (Scandinavian Type, FBB)	58
1.3	White lined chipboard (WLC)	60
1.4	Solid bleached board (SBS)	60
1.5	Solid unbleached board (SUS)	61
1.6	Liquid packaging board (LPB)	62
2	**Containerboards**	**64**
2.1	Linerboard	66
2.2	Corrugating medium	70
3	**Special boards**	**70**
3.1	Wallpaper base	70
3.2	Core board	70
3.3	Plaster board	71
3.4	Some other specialty paperboards	71
	References	72

CHAPTER 2

Ari Kiviranta

Paperboard grades

There is a wide variety of paperboards on the market. Paperboards are classified into three categories: cartonboards, containerboards, and specialty boards. The classification used in this chapter is shown in Fig. 1.

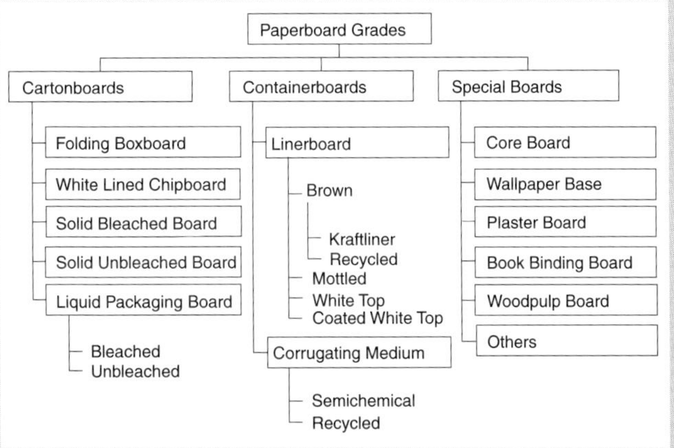

There is no clear definition how paperboard differs from paper. However, there are certain things that are common for most paperboard grades.

Figure 1. Classification of paperboard grades.

Usually basis weight of paperboards is higher than 150 g/m^2. There are exceptions like many linerboard and corrugating medium grades that can have a basis weight lower than 100 g/m^2. Also, most of the paperboard grades are multi-ply products like folding boxboard and liquid packaging board. Corrugating medium is an exception also here since it is very often a single-ply product.

Paperboards are very often used for packaging, but again there are exceptions like plasterboard. Because of their function in the packages, strength properties are most often very important for paperboards.

Cartonboards are mainly used for consumer product packaging for products such as food, cigarettes, milk, and pharmaceuticals. World production of cartonboards totalled 32 million tons in 1998.

Containerboards are a very big market in paperboard industry. Corrugated boxes are used in many packaging applications starting from simple transportation containers and ending with multicolor printed display containers for stores. In 1996 the world total for corrugating materials, including all paperboard types for corrugated boxes, was almost 80 million t. Figure 2 shows the distribution of corrugating materials production throughout the world.

CHAPTER 2

Paperboard has to compete with other packaging materials such as plastics. Various types of paperboard make very good and durable packages that are also recyclable. Old corrugated containers, for instance, are very efficiently used as raw materials for linerboard and corrugating medium. The fact that paperboard products are environmentally friendly and recyclable makes them very competitive against many other packaging materials.

1 Cartonboards

Cartonboards are divided into various subgrades. Table 1 shows typical classifications. The abbreviations shown in the table are used later in the text to indicate various cartonboard grades. This classification is made based mainly on the raw material used.

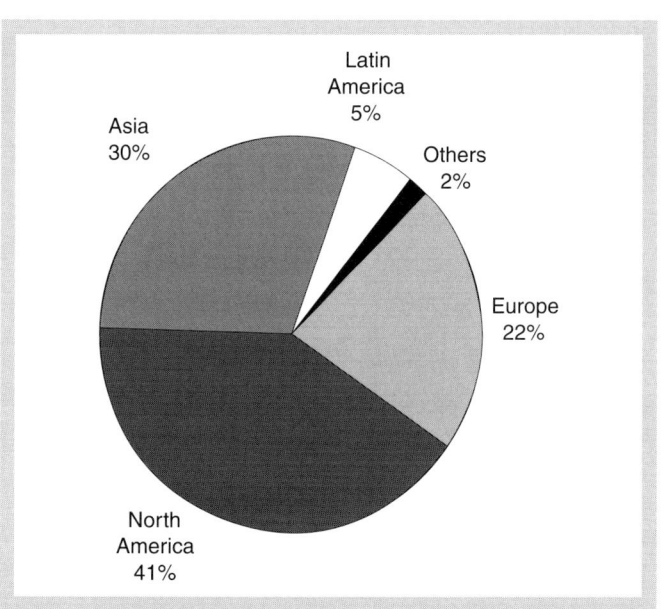

Figure 2. Distribution of corrugating materials production in the world[1]. Total production was 79.5 million t in 1996.

Table 1. Cartonboard grades.

Grade	Abbreviation
Folding boxboard (Scandinavian type)	FBB
White lined chipboard	WLC
Solid bleached (sulfate) board	SBS
Solid unbleached (sulfate) board	SUS
Liquid packaging board	LPB

For good printing properties, most of the cartonboard grades are pigment coated. The printed side is often double coated; in some cases discussed later, the top side can be triple coated. There are also certain applications in which rough appearance is needed and uncoated cartonboard is used.

1.1 Boxmaking process

Each end use sets its own demands on the properties of cartonboard. There are, however, some basic requirements common to most of the end uses. The requirements can be divided into mechanical and functional characteristics, purity and cleanliness, and visual characteristics. This is discussed in detail in volume 12, *Paper and Paperboard Converting,* of this series.

1.1.1 Mechanical requirements

Cartonboard is used to produce folding carton. A basic requirement is a certain level of mechanical strength and stiffness. The cartons may be stacked on top of each other, and therefore sufficient compression strength is required. Also, for good looking cartons, the carton must not bulge. Bending stiffness, and especially cross machine direction bending stiffness, is critical. Bending stiffness is affected mainly by thickness of paperboard and modulus of elasticity given by the raw materials in it. The optimum structure is that the middle ply is very bulky and the top and back plies have high modulus of elasticity. The lower the basis weight that can be achieved at given stiffness, the better the yield is. Then more area can be produced from the same weight.

In some industries, like cigarette packaging, the packaging lines are of very high speed. This sets high requirements for paperboard runnability in the converting machine. For good runnability, the paperboard used must have a certain, and consistent, curl.

When the paperboard is creased, there are tensile, compression, and shear forces acting on the board. Cracking of board surface should be avoided. There are a lot of creasing parameters that can be used to control and eliminate cracking. Cracking tendency is also affected by board structure. To minimize cracking tendency, the stretch at break should be as big as possible for the top ply. Also z-directional strength is important. If z-strength is too low, the board can delaminate in printing. On the other hand, too high of a z-strength can cause cracking since the stretch is too big for the top ply, if the middle ply does not delaminate in creasing.

After printing, cutting and creasing, and embossing, the cartons are glued. The paperboard must have sufficient porosity for good gluing results.

1.1.2 Purity and cleanliness

Purity and cleanliness requirements are high especially if food items are packed. In that case, microbiological purity is also vital. There might be some microbes in the paperboard that do not cause health problems, but they might cause problems with odor and taint. Virgin fiber-based paperboards do not usually have problems with microbiological purity, but recycled fiber-based product might have, due to the variability of the fiber and fiber source. Therefore there are some food packaging applications, in which the paperboard is in direct contact with food, where recycled fiber based paperboard cannot be used.

As mentioned, in most cases microbes do not provide a health risk, but they still can destroy the product by causing odor and taint problems. Examples of products that are very sensitive to odor and taint are chocolate and cigarettes.

1.1.3 Printing properties

Almost all cartons contain printing. The package design defines the printing quality requirements. There are very demanding print jobs made on cartonboards, like packages for luxury items such as cosmetics.

CHAPTER 2

The main printing method for cartonboards is sheet-fed offset. Also here mechanical stability and consistent quality are essential. For a good printing result, a certain level of macro and microscale smoothness is required. Offset printing introduces a lot of stress on paperboard surface; therefore, good z-directional strength, both Scott Bond and IGT surface strength, is needed. In rotogravure printing smoothness is very important for good printing results. Here z-strength is not as critical as with offset printing.

1.1.4 Summary of boxboard requirements

Table 2 summarizes requirements set on various paperboards.

Table 2. Examples of packaged products, their special requirements for the carton and typical cartonboard grades (from volume 12, *Paper and Paperboard Converting* of this series).

Product	Special requirements	Typical cartonboard grade
Direct food	Purity, cleanliness, runnability	FBB
Frozen food	Strength, barrier, purity, cleanliness, runnability	SBS, SUS
Indirect food	Runnability	WLC
Confectionery	Attractive appearance, purity, cleanliness, odor and taint free	FBB, SBS
Bottle carriers	Strength	SUS
Cosmetics, toiletries	Attractive appearance	FBB, SBS
Cigarettes, tobacco	Runnability, odor and taint free, appearance	SBS, FBB
Pharmaceuticals	Identification, runnability	FBB, WLC
Detergents	Strength, runnability	WLC, SUS
Household durables, hobby items	Strength	WLC
Textiles, clothing, footwear	Appearance	WLC, FBB
Toys, games	Strength, purity	WLC, SUS
Paper products	Appearance, runnability	WLC
Milk, juices	Runnability, cleanliness, purity, strength	LPB

1.2 Folding boxboard (Scandinavian Type, FBB)

Scandinavian type folding boxboard is used for various packaging applications. Typical products to be packed in folding boxboard include cosmetics, cigarettes, pharmaceuticals, confectionery items, and food. Some grades of folding boxboard are used for postcards and book covers. Basis weight

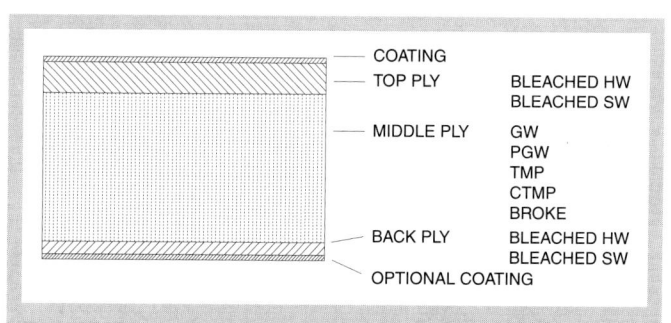

Figure 3. Typical structure of Scandinavian type folding boxboard.

for folding boxboard is 160–450 g/m². Folding boxboard is delivered to the customer either as rolls or already sheeted, depending on the printing method.

Folding boxboard is typically made of three, or in some cases four, plies (Fig. 3). Top and back plies are made of bleached chemical pulp. Mechanical pulp (groundwood, pressure groundwood, TMP, or CTMP) and machine broke are used for the middle ply or plies. Mechanical pulp is used in the middle ply to give the highest possible bulk. Also, basis weight of the top and back plies are minimized for high bulk and also for lower raw material cost.

Folding boxboard can be surface sized. Depending on the end use, board can be uncoated, single coated on the top side, double coated on the top side, or double coated on the top side and single coated on the back side.

Since folding boxboard is used for demanding packaging, printing properties of the top surface are very important. Smoothness is usually measured as Parker Print Surf (PPS) roughness. Common printing process for folding boxboard is sheet-fed offset. Therefore good surface strength is required to prevent delamination, if tacky printing colors are used. This is measured as IGT surface strength.

The most important mechanical properties for folding boxboard include thickness, bending stiffness, and z-directional strength, usually measured as Scott Bond. Higher caliper results in better bending stiffness. On the other hand, because smoothness is important, the most critical property pair for folding boxboard is smoothness–bending stiffness. For this reason, smoothness cannot be made only with calendering. Folding boxboard machines typically have an MG dryer (Yankee cylinder) for improved top side smoothness. MG gives a smooth surface and also closes the surface structure very efficiently for coating process.

The forming section of a modern folding boxboard machine provides separate forming of individual plies with fourdriniers. Middle-ply fourdrinier is equipped with a top

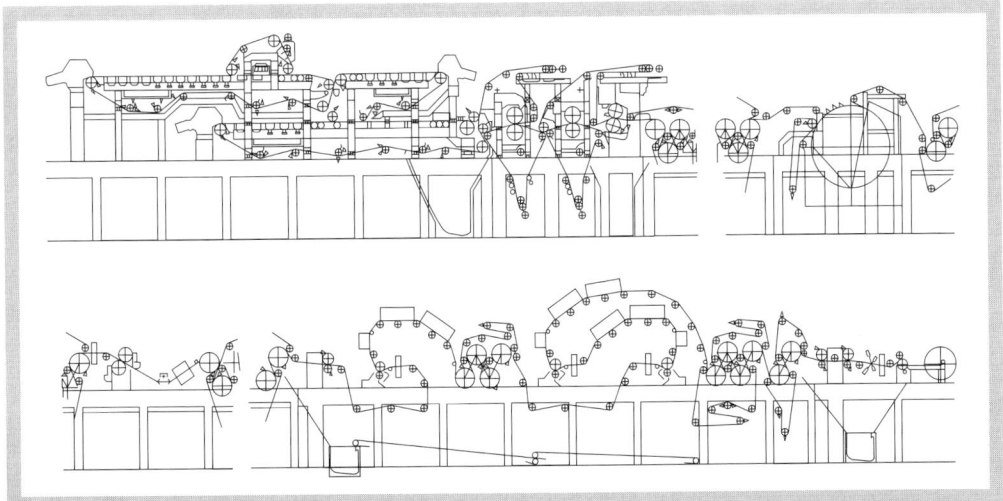

Figure 4. Folding boxboard machine with a three-ply forming section, MG dryer, size press, and three coating stations.

dewatering unit to improve formation and increase drainage capacity. The press section usually has two or three straight-through presses of which one is a shoe press or a large diameter roll for maximum bulk. The pre-dryer section is followed by an MG, an after-dryer section, a film-type size press, calendering, and on-machine coating. This makes folding boxboard machines one of the most complicated paper or board machines. Coating sections for FBB and other coated board grades are described in volume 11, *Pigment Coating and Surface Sizing of Paper*, of this series. Figure 4 shows a generic folding boxboard machine.

1.3 White lined chipboard (WLC)

White lined chipboard is used for similar end uses as FBB; typical basis weight is 200–450 g/m^2. The structure is very similar, although different furnishes are used (Fig. 5). Top ply is usually made of bleached chemical pulp; also deinked office waste or white ledger may be used. Typically between the top and middle plies is an undertop ply. Deinked pulp, white ledger, or mechanical pulp is used

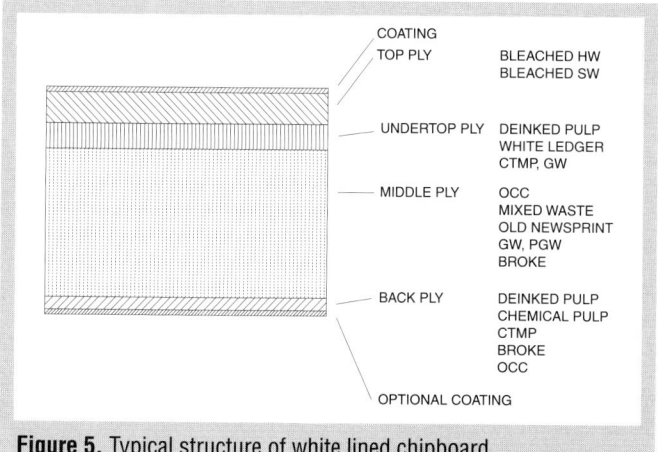

Figure 5. Typical structure of white lined chipboard.

here. The idea of having an undertop ply is to minimize the amount of expensive top ply pulp. This is because the middle ply is made of recycled fiber that has low brightness. Middle ply furnish can be a mixture of old corrugated containers, old newsprint, and mixed waste, for instance. Furnish composition of the middle ply varies a lot between the mills, depending on the availability of various pulp types in the area. Typical back ply furnish is deinked pulp, or bleached chemical pulp.

Because of the various recycled fiber types in the middle ply, white lined chipboard cannot be used for certain food packaging applications like chocolate packaging. However, together with a protective plastic bag between the product and the package, groceries like cereals can be packed in white lined chipboard. As with FBB, WLC is also available in rolls or as sheets.

Same board properties are important for white lined chipboard as for folding boxboard. Also the board machines for white lined chipboard and folding board are very similar. The greatest difference is obviously in the stock preparation area.

1.4 Solid bleached board (SBS)

SBS (Solid bleached sulfate) belongs to the same product family with Scandinavian type folding boxboard and white lined chipboard, since it is used for similar applications. For certain products, like chocolate and cigarette packaging, SBS is preferred since it does not usually cause problems with odor and taint. Also SBS is available in rolls or as sheets.

The main furnish component to produce SBS is bleached hardwood sulfate pulp; bleached softwood sulfate is used as well. Hardwood pulp results in better formation and gives better printing properties.

In North America, a modern forming section of an SBS machine consists of a fourdrinier and top dewatering unit; as mentioned before, SBS is usually a single-ply product (Fig. 6). However, it is possible to produce SBS with multi-ply techniques. A triple fourdrinier concept with a top dewatering unit on the middle-ply can be used, for instance. In that case, furnish composition for each ply can be optimized individually for bulk, stiffness, and printing properties (Fig. 7).

SBS is very often coated. Board is calendered with so-called "wet stack" before coating. Water is fed into two nips in a wet stack. The idea is to soften the top fibers and improve smoothness. Wet stack is very effective in improving smoothness; however, bulk is reduced simultaneously. Also, with low basis weight, wet stack can cause runnability problems.

Figure 6. Single-ply solid bleached board.

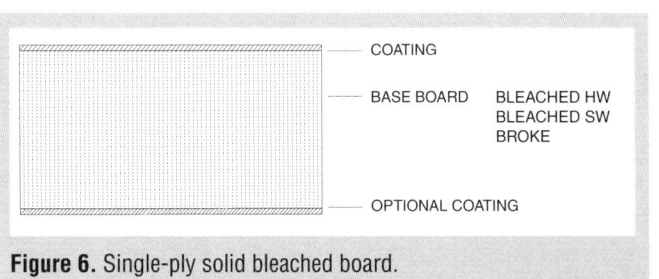

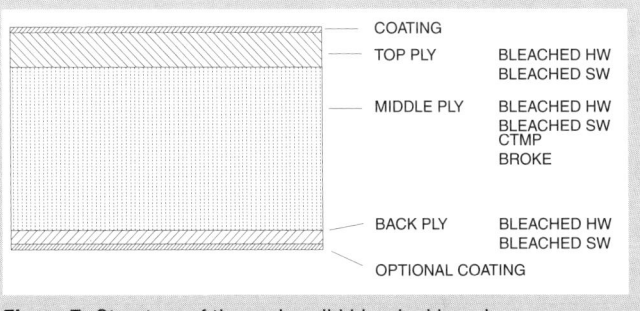

Figure 7. Structure of three-ply solid bleached board.

1.5 Solid unbleached board (SUS)

Solid unbleached sulfate (SUS) is typically a multi-ply product and is used for consumer packages of beverages, etc. The basis weight of carrier board can be up to 500 g/m^2. SUS is typically sold as rolls.

Solid unbleached board is typically made of two or three plies (Fig. 8). All plies of the base board are made of unbleached furnish. Since the top ply should give a good surface for coating, hardwood furnish is typically used for it.

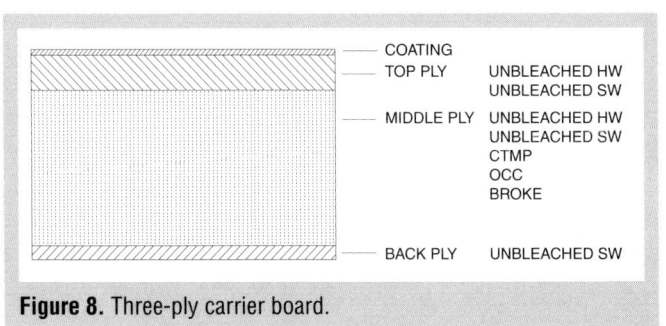

Figure 8. Three-ply carrier board.

CHAPTER 2

The middle ply (or in the case of two plies, the back ply) consists of unbleached softwood kraft and broke. Also recycled fiber, like old corrugated containers, can be used in the middle ply. Back ply is usually made of unbleached softwood kraft.

SUS can be surface sized. Board is often calendered with a wet stack (as SBS) before coating. A unique feature in SUS manufacturing is that the unbleached base sheet can be coated. This is accomplished with three coating layers so that the middle coating is traditionally done by an air-knife coater. Also, titanium dioxide is used as a pigment in the coating to give good opacity of the coating and therefore good coverage.

As for any cartonboard, stiffness is a very important property for SUS. Good stiffness is obtained by the proper degree of refining of the back and top plies, surface sizing, and maximizing bulk. To prevent any cracking of the beverage package, high tear strength is important. Also, a certain level in z-directional strength is required.

Since SUS is used for relatively demanding printing jobs, printing properties of the top surface are very important. Smoothness is usually measured as Parker Print Surf (PPS) roughness.

The forming section of a modern SUS machine has separate forming of individual plies with fourdriniers. Middle-ply fourdrinier is equipped with a top dewatering unit to improve formation and increase drainage capacity. The press section usually has two or three straight-through presses of which one is a shoe press for maximum bulk. The machine is also equipped with a film-type size press, a wet-stack calender, and on-machine coating. Figure 9 shows an example of a carrier board machine.

1.6 Liquid packaging board (LPB)

There are various kinds of liquid packaging boards on the market. Liquid packaging board is used for various liquid packaging applications, of which the most common is

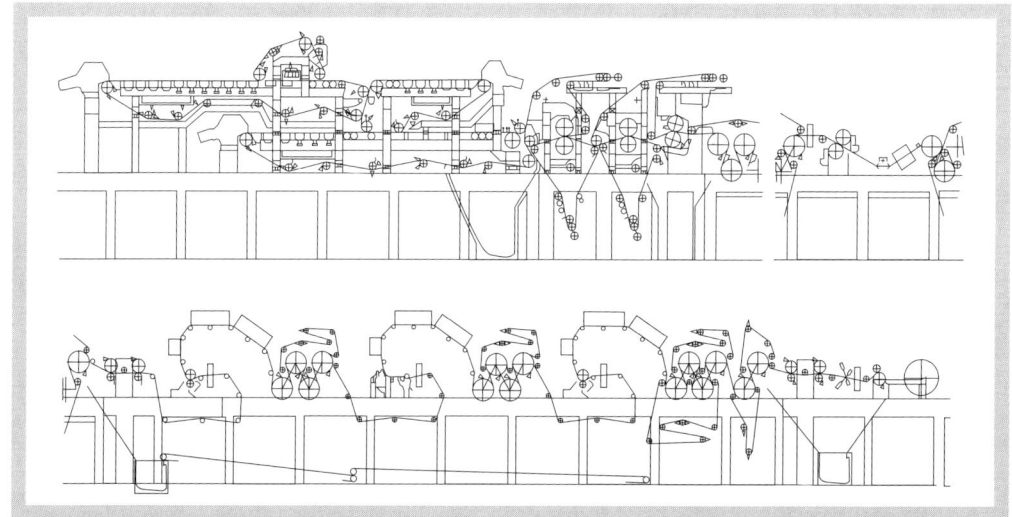

Figure 9. Carrier board machine with a three-ply forming section, a press section with a smoothing press, a size press, two calender nips before coating, a coating section, and two-nip calendering after coating.

Paperboard grades

milk and juice packaging. There are basically two types of liquid packages. Two-sided low-density polyethylene-coated paperboard is used for pasteurized milk products. Long shelf life dairy products require high barrier coated paperboard. There are also many non-food applications that require a high barrier coating to prevent flavors from exiting the package. This method of coating is discussed in Volume 12 of this book series.

LPB is used for two package types: brick and gable top. Brick board is delivered to dairies in rolls containing the board in flat form or in tubular form with the longitudinal seal already made. The unique feature of the packaging process is that filling and forming of the package are made in one stage under a liquid surface. This means that there will not be any gas in the package. Also, the contents support the carton; therefore, lower basis weight can be used. Brick type packages are often aseptic.

LPB for gable top packages, like one-liter milk cartons, is delivered in blank form with the side seal already made. The bottom seal is made after the carton has been assembled, and then liquid is added, after which the top seal is made. Higher basis weight LPB is used for gable top packages than that used for bricks.

Because LPB is used for milk and juice packaging, purity and cleanliness of the product are of utmost importance. This means that only virgin fiber can be used for LPB. For good package, high paperboard stiffness is needed. This has led first to two-ply, and finally three- (or even more) ply structure. Pulp giving high modulus of elasticity is used for the top and back plies; pulp giving maximum bulk is used for the middle ply.

Since LPB goes through a converting operation in which various layers of plastic and/or metal is added, LPB is delivered from the board mill to converters in roll form.

Z-directional strength is needed to prevent delamination and leakages. Printing surface requirements vary a lot based on the end use. Gable top milk cartons are often printed with flexo printing; therefore, the surface quality needed is not very critical. On the other hand, there are brick cartons that are printed with rotogravure printing, thus needing excellent small-scale smoothness usually measured as PPS.

Figure 10. Two-ply liquid packaging board.

LPB is a multi-ply product that can be coated or uncoated depending on the end use. Also, LPB can be made of unbleached or bleached chemical pulp. Figures 10 and 11 show examples of LPB structures. The idea in LPB structure is the same as with FBB: Use a high

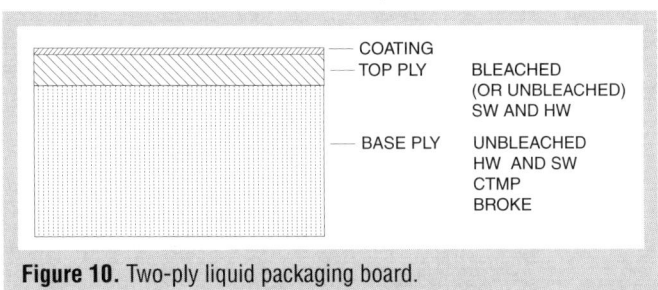

Figure 11. Three-ply liquid packaging board.

CHAPTER 2

modulus of elasticity pulp in the outer plies and bulky pulp in the middle ply to maximize stiffness. The latest trend in LPB has been the use of CTMP, which has clearly higher bulk than chemical pulp, in the middle ply to increase bulk, thus making it possible to use a lower basis weight for a given stiffness. LPB is strongly wet strength sized.

As multi-ply structures are needed, the forming section of a LPB machine can be very complicated. There are machines with a long base fourdrinier that form the top ply. Middle plies and the back ply are formed on the top ply with secondary headboxes and top dewatering units. The number of plies can be up to five. Also a multi-fourdrinier concept with three plies is used. In that case, the middle-ply fourdrinier is equipped with a top dewatering unit. The most exotic technique in LPB manufacture is to use high-consistency forming for the middle ply. High-consistency forming gives higher bulk, especially with chemical pulp, than conventional low-consistency forming. The drawback with high-consistency forming is worse formation with a grainy appearance; therefore, it is not very suitable in its present state for high-quality rotogravure LPB.

The press section usually has a shoe press to maximize bulk at a given dryness after the press section. The machine is also equipped with a film-type size press. Depending on the LPB grade, the machine can be with or without on-machine coating. On older machines, wet stack is still used for calendering; on modern machines, the wet stack is replaced with hot soft calendering.

2 Containerboards

Containerboards – linerboard and corrugating medium – are used to produce corrugated board corrugated containers. Corrugated board as such is an old packaging material, but it is still very competitive in many packaging applications.

Corrugated boards can be divided into following categories:

- Single-faced corrugated
- Single-wall corrugated
- Double-wall corrugated
- Triple-wall corrugated.

Single-faced corrugated board (Fig. 12) has one liner and one layer of corrugated material glued together. This is usually not used for corrugated boxes, but as protecting material. *Single-wall corrugated board* is basic raw material for corrugated boxes. The corrugated material is sandwiched between two liners. *Double-wall corrugated board* consists of two layers of corrugated material and three liners, one of which is between the two corrugated layers. Double-wall corrugated is used when very rigid and strong material is needed. *Triple-wall corrugated board* consisting of four liners and three layers of corrugated material offers even stronger and more rigid structure.

Basis weight of the corrugated board depends on the end use and the protection needed. For high-strength applications, triple-wall corrugated board with high basis weight linerboard and corrugating medium is used. Micro-flute corrugated board represents the other end of the spectrum, micro-flute is increasing its share in consumer packaging market in which folding boxboard and SBS are traditionally used.

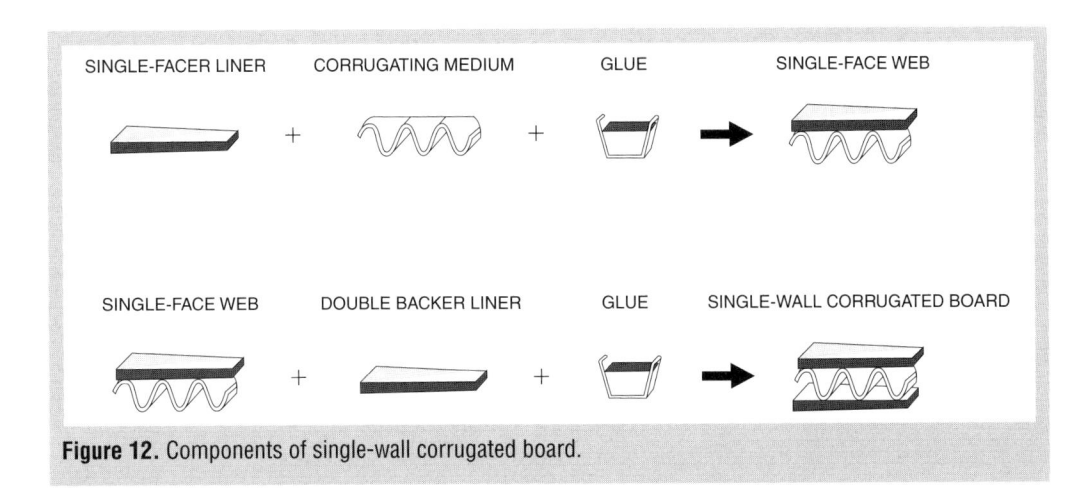

Figure 12. Components of single-wall corrugated board.

1. PREHEATING
2. CORRUGATING ROLLS
3. GLUING
4. BRIDGE
5. HOT PLATES
6. PRESSURE ROLLS
7. SLITING AND SCORING
8. CUTOFF

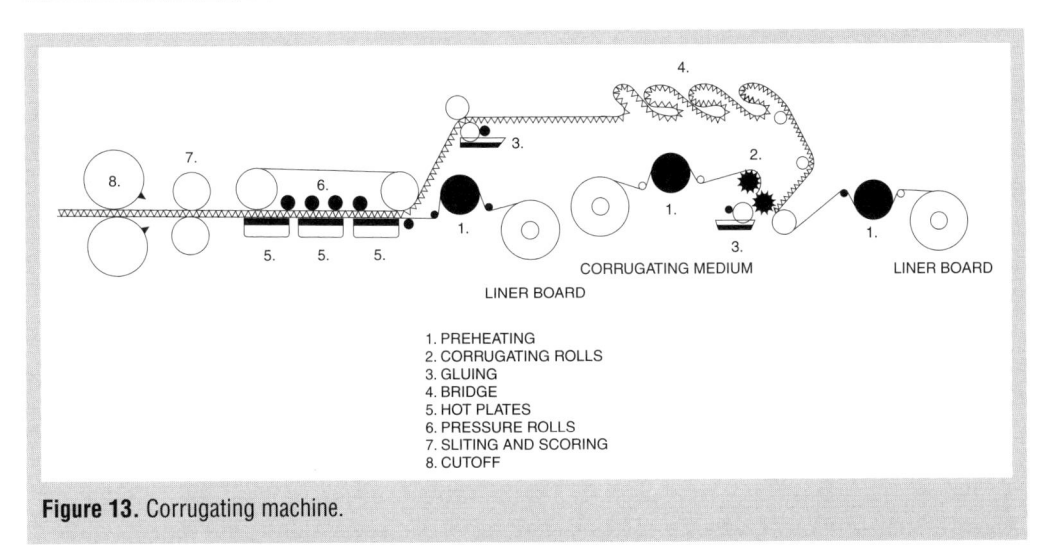

Figure 13. Corrugating machine.

Production of corrugated board starts with corrugation process. To soften lignin and hemicelluloses in corrugating medium for easier corrugating, the web is preheated. After preheating, the web is presteamed to further increase temperature and to moisten the web. The final temperature of the web just before corrugation is slightly below 100°C.

In corrugation, (Fig. 13) the web is run between two heated corrugating rolls. The rolls are steam heated to 160°C–190°C. Afterward, corrugating glue is applied on the tips of the flutes. Liner is pressed against the corrugating medium with a pressure roll to form single-faced corrugating board. This is followed by a short bufferstock called a "bridge" that acts as a preparatory step for the second gluing. Also, when the corrugated board is preheated, the first glue dries. Then glue is applied on the other side of corrugated medium, and the second liner is added to make single-wall corrugated board.

For good runnability on a corrugating machine, an even moisture profile of both liners and corrugating medium is needed. Also, the moisture level should be 8%–9%.

CHAPTER 2

Gluing is a critical stage in corrugated board manufacture. The gluing result is affected by many factors including glue, the temperature used in the corrugating machine, operating conditions, and paperboard properties. There is no good test for gluability. Still, it is believed that porosity, water absorption properties, and the moisture of both linerboard and corrugating medium have an effect on gluability.

The corrugated box should protect its content as well as possible. Therefore strength properties of corrugated board are very important. Strength is typically measured with the edge crush test (ECT), flat crush test (FCT), and bursting strength. The strength of gluing is tested with the pin adhesion test (PAT).

The creased and cut corrugated board should be sufficiently flat for good runnability on conversion and packaging machines. However, due to paperboard structure and production methods, there might be some warp. So-called "twist warp" (the opposite corners of the board are turned either up or down) can be caused by differences in the fiber orientation angles of the liners. Therefore an even fiber orientation profile on a linerboard machine can be important.

When corrugated board is creased and cut, some cracking close to the crease or cut may appear. This is well correlated with tear strength of linerboard. It is well known that tear strength highly depends on the fiber length of the furnish. Therefore, virgin fiber-based linerboard has a clear advantage over recycled fiber based linerboard in this respect.

2.1 Linerboard

Linerboard is used as liner for corrugated board. The typical basis weight range for linerboard is 125–350 g/m^2, although basis weights clearly below 100 g/m^2 are adequate for small boxes.

There are various grades of linerboard on the market. Linerboard is almost always a two-ply product consisting of top and base plies. Linerboard is manufactured with various top plies and from various raw materials. Both virgin (typically softwood, in some cases hardwood or sawdust pulp) and recycled fiber are used in linerboard manufacture. Linerboard manufactured mainly from virgin fibers is usually called kraftliner, whereas recycled fiber containing linerboard is called testliner. However, during the recent years, almost all new linerboard capacity is based on recycled fiber. In North America, old corrugated containers (OCC) are used as recycled fibers; in Central Europe and Asia, mixed waste is also utilized. If virgin softwood kraft pulp is used, the pulp for base ply is cooked to a high yield and then refined slightly. Top ply pulp is cooked to a lower kappa number and also refined more that the base ply furnish. Linerboard is resin sized for moisture resistance. If low-quality recycled fiber is used as raw material, a size press for surface sizing is needed to produce sufficient strength properties.

In brown linerboard (Fig. 14), both top and base plies are unbleached.

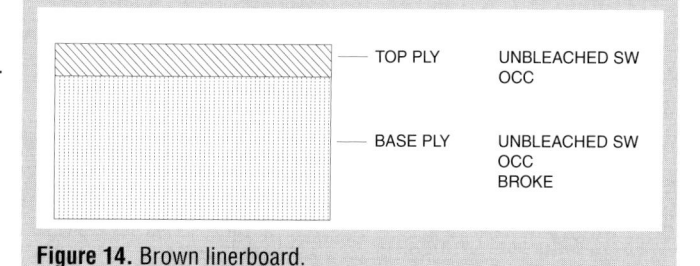

Figure 14. Brown linerboard.

Paperboard grades

Depending on the machine design, the basis weight split between the top and base plies is typically 30/70. Brown linerboard is the simplest product of linerboards, since the requirements for printing properties are not very high.

Mottled linerboard has a top ply made of bleached pulp. Top ply basis weight is roughly 40 g/m^2. The low top ply basis weight, as such, gives the mottled linerboard its flocky appearance. To enhance the appearance, the top ply is run at higher headbox consistency than other linerboard grades to make the top ply formation worse. Also, jet-wire speed difference is typically increased to make formation worse.

White top linerboard (Fig. 15) is used for more demanding printing jobs; therefore, smoothness and appearance of the top surface are very important. Basis weight for the top ply is 70–80 g/m^2, and bleached chemical pulp is used in the top ply. If excellent formation is to be achieved, hardwood pulp should be the main component of the top ply. Filler is often used in the top ply to improve opacity, and therefore appearance, of the top side. Here the top ply formation should be as good as possible for visual appearance.

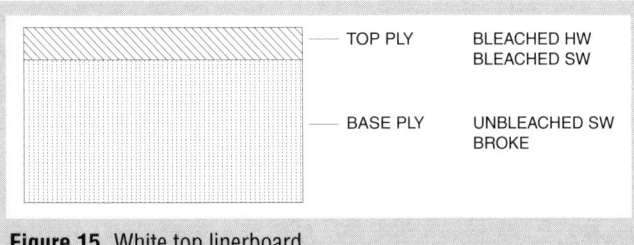

Figure 15. White top linerboard.

Coated white top linerboard is the most demanding product in the linerboard family. Top ply is made of bleached pulp, and its basis weight is the same as for white top linerboard. Board can be coated either on machine or off machine. Typical coating is a single coating with a blade coater. Coated white top linerboard is used for demanding corrugated board applications, like for containers that are used in a display in stores.

Testliner is produced especially in Central Europe and in Asia. Various furnishes are used in testliner (Fig. 16). Because of the mixed waste used, a four-ply structure is often preferred for two reasons. First, by using four plies, the linerboard properties can be optimized better than with a two-ply structure. Second, since the drainage resistance of the mixed waste is very high, it would be difficult to dewater one heavy base ply on the wire section. By splitting the base ply into 2–3 plies, drainage on the forming section becomes a lot easier.

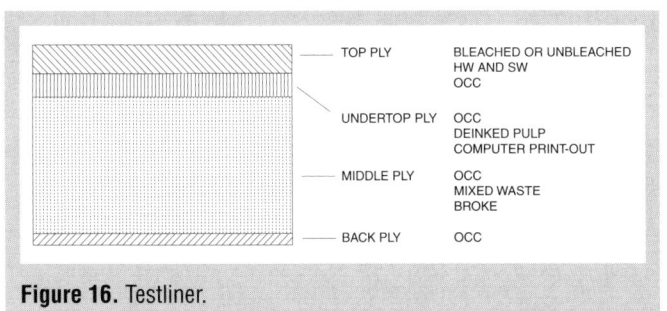

Figure 16. Testliner.

Burst and cross machine directional compression strength are the most important strength properties for linerboard. Burst is measured with Mullen tester, as compression strength is measured by either the ring crush test or short column test (SCT, sometimes also called STFI test). Although the printing process for linerboard is not very demanding (especially for brown linerboard), certain smoothness is required. Smoothness is

CHAPTER 2

measured by Bendtsen or Sheffield method; also the Emveco tester is used. The top side friction coefficient is important because the stacked corrugated containers should not slide. This property is measured as a slide angle.

There are various designs for the forming sections of a modern linerboard machine. The basic structure of the board is two plies. This can be made with two fourdriniers (Fig. 17), a fourdrinier and a gap former (Fig. 18), or with a single gap former equipped with a multichannel headbox. Separate forming of plies is needed to produce mottled, white top, and coated white top linerboard. Therefore the most common forming section for those grades is a base fourdrinier and an on-top fourdrinier that can be equipped with a top dewatering unit. It is also possible that the top ply unit is a gap former or that both units are gap formers. The same wire sections are used also for brown linerboard. In some cases, when mixing of the white waters is not critical, like with brown linerboard simultaneous forming with a multichannel headbox may be used. Currently, the biggest challenge for gap formers in linerboard manufacture is to achieve the highest possible cross machine direction strength properties as produced by a fourdrinier machine.

The modern press section of a linerboard machine has at least one shoe press; in some cases, the press section consists of two shoe presses. Because linerboard is usually not a bulk-sensitive grade, the highest press loads allowed by the press design are used to maximize production.

Dryer section is simple, consisting of single and/or two tier dryer groups. When low-quality recycled fiber is used as raw material, the machine is equipped with a size press to improve strength properties. The size press can be of a conventional puddle-type or a film-type size press. The board is finished with a hot calender nip that can be either soft or hard.

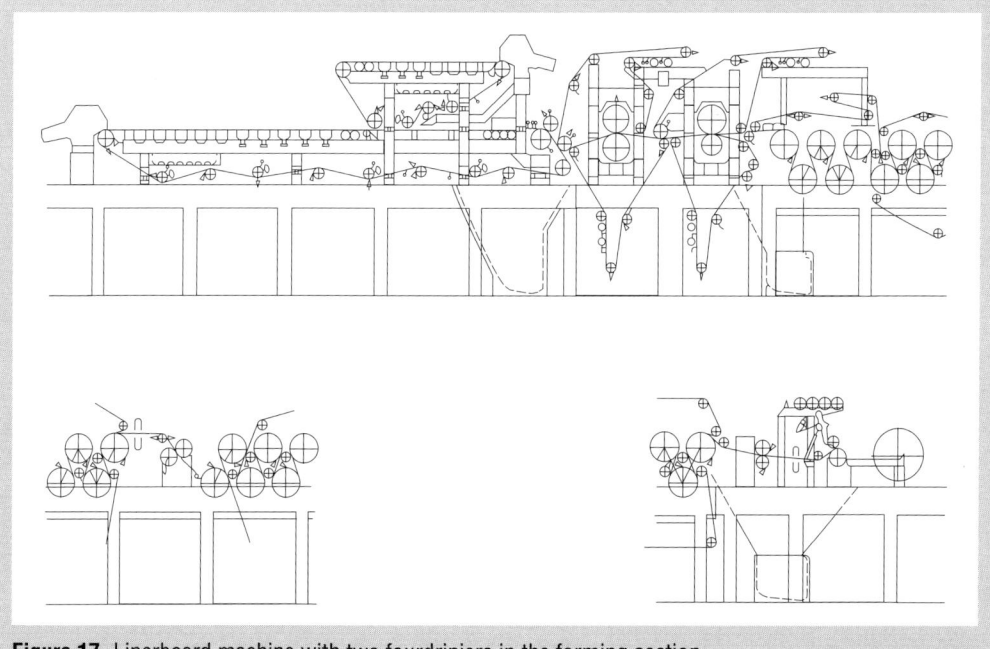

Figure 17. Linerboard machine with two fourdriniers in the forming section.

Paperboard grades

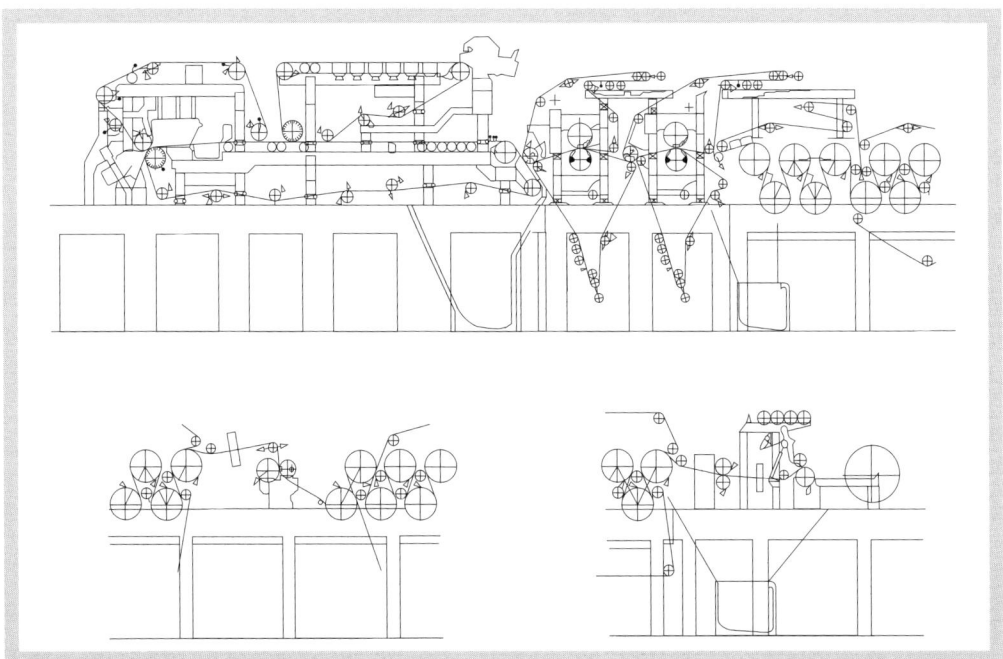

Figure 18. Linerboard machine with a gap former and a foudrinier in the forming section.

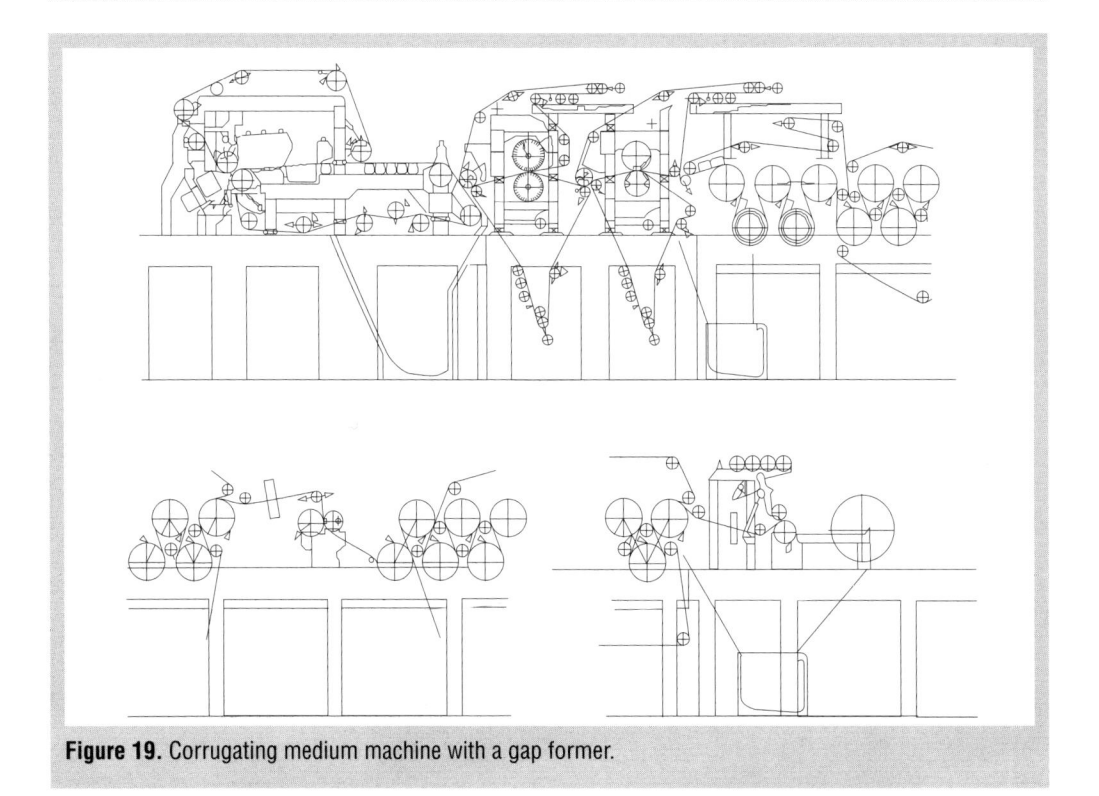

Figure 19. Corrugating medium machine with a gap former.

CHAPTER 2

2.2 Corrugating medium

Corrugating medium is used in corrugated board as the "wave" between the two liners. The basis weight range for corrugating medium is 112–180 g/m^2.

Both semichemical pulp (if this is the only raw material, corrugating medium is often called "fluting") and recycled fiber are used as raw material for corrugating medium. Because corrugating medium is a bulk product, price is very important for both raw materials. Corrugating medium is a single-ply product. If the raw material is very low-quality mixed waste, surface sizing can be used to improve strength properties.

Compression strength of corrugating medium is an important strength property, which is measured with the Concora test.

A modern corrugating medium machine has a gap former forming section (Fig. 19), although there are numerous fourdrinier machines in use. Press section can be equipped with a shoe press to maximize production. Dryer section is simple, consisting of single or double tier sections. Corrugating medium is not calendered.

3 Special boards

3.1 Wallpaper base

There are many grades of wallpaper base, and one of them is a close relative to folding boxboard. Wallpaper base can be produced from chemical pulp, but also mechanical pulp may be used in a multi-ply structure. Wallpaper base is delivered as rolls. A more detailed description of the use and requirements for wallpaper base can be found in the specialty paper section of this book.

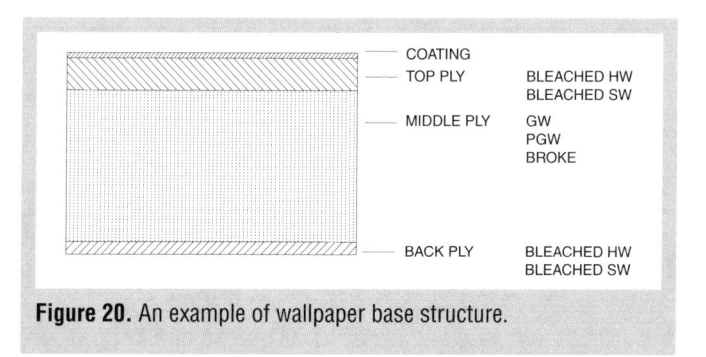

Figure 20. An example of wallpaper base structure.

If mechanical pulp is used, wallpaper base has a two- or three-ply structure (Fig. 20). Mechanical pulp is used in the middle (or base) ply to give good bulk. Top ply is made of chemical pulp for good printing properties. The basis weight range for multi-ply wallpaper base is 120–200 g/m^2. Wet end starch is used; wallpaper base can also be sized for wet strength.

Multi-ply wallpaper base is typically produced on multi-fourdrinier machines. Press and dryer sections are conventional. Surface characteristics are improved in some cases by an MG-dryer; also soft calenders are used. If wallpaper base is coated, on-line coating is typically used.

3.2 Core board

Core board is used to make cores for instance for paper rolls. Recycled fiber is used as raw material. Core board can be either a single- or multi-ply product.

The most important strength property for core board is z-directional strength, measured as Scott Bond. This is to prevent delamination of the core during winding.

Core board is produced with a variety of machines. Some machines have a cylinder mold forming section, but also fourdrinier forming is used. Press section is designed to maximize production, even though in some cases it is important to maintain high bulk. Then a shoe press can be used. Dryer sections typically consist of double felted dryer sections.

3.3 Plaster board

Plaster board is used as liners for gypsum board, which is a popular wall covering material. When producing the gypsum board, wet gypsum is spread on the first liner. The second liner is then set on the wet gypsum. The typical basis weight for plaster board is 200–300 g/m².

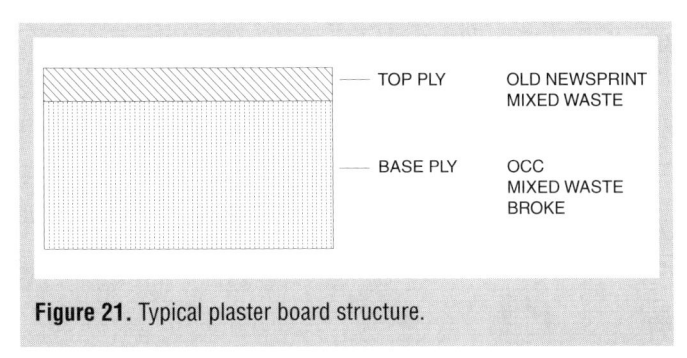

Figure 21. Typical plaster board structure.

Various recycled fiber types are used in plaster board, which is a multi-ply product (Fig. 21). Top ply can be made of old newsprint, for instance. Base ply furnish can be a mixture of OCC, old newsprint, and broke.

High machine directional tensile strength is important for plaster board because of the handling of gypsum board (the board is lifted from the shorter end when a wall is built). Also, a certain porosity level is needed for proper adhesion.

There are various types of machines for plaster board. Multi-fourdrinier forming sections are common; the number of plies can vary between 2 and 4. A modern press section has a shoe press. Conventional two tier drying is used.

3.4 Some other specialty paperboards

There are many specialty paperboards on the market, although their production is very small compared to the paperboard grades discussed before. *Book binding board* is used in covers for hard-cover books. The most important property is stiffness. It is typically produced on small cylinder mold machines from recycled fiber. Book binding board can be so thick and stiff that it cannot be reeled after drying, so it has to be cut into sheets right on the paper machine.

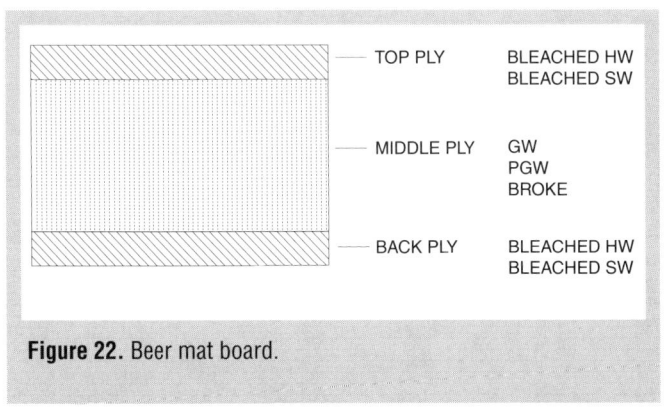

Figure 22. Beer mat board.

Wood pulp board and *beer mat board* are small paperboard grades. As the name says, beer mat board is used in restaurants under beer glasses. It is usually printed. Figure 22 shows the typical structure of beer mat board.

CHAPTER 2

References

1. Anon., PPI's International Fact & Price Book, Miller Freeman Inc., San Francisco, 1997.

CHAPTER 3

Tissue

1	**Tissue markets**	**75**
2	**Raw materials for tissue production**	**77**
2.1	Pulps	77
	2.1.1 Virgin fiber pulp	77
	2.1.2 Recycled fiber pulp	78
	2.1.3 Pulp mix	78
3	**Uses of tissue**	**79**
3.1	Hygiene products	79
	3.1.1 Bathroom tissue	79
	3.1.2 Kitchen towels	80
	3.1.3 Paper towels and industrial wipes	80
	3.1.4 Facial tissue and handkerchiefs	81
	3.1.5 Table napkins	81
3.2	Other tissue products	82
3.3	Quality requirements	82
	3.3.1 Basis weight	82
	3.3.2 Absorbency	82
	3.3.3 Softness	83
	3.3.4 Thickness and bulk	83
	3.3.5 Tensile strength	83
	3.3.6 Brightness, whiteness	83
	3.3.7 Crepe ratio, stretch	84
	3.3.8 Appearance	84
	3.3.9 Quality requirements for converting	84
4	**Special features of the tissue production process**	**84**
4.1	Multilayer headbox	85
4.2	Tissue forming sections	86
	4.2.1 Fourdrinier machines	86
	4.2.2 Suction breast roll machines	86
	4.2.3 Twin-wire machines	87
	4.2.4 Crescent former machines	87
4.3	Drying section, creping	88
4.4	Through-air drying system (TAD)	89
5	**Tissue finishing**	**89**

CHAPTER 3

6	**Tissue converting**	**90**
6.1	Embossing	90
	6.1.1 Traditional embossing	90
	6.1.2 Nested embossing	91
	6.1.3 Foot-to-foot embossing	91
6.2	Printing	91
6.3	Perforation	91
6.4	Winding and tail sealing or folding	91
6.5	Log sawing	91
6.6	Packaging	91
7	**Environmental impact of tissue manufacture**	**92**
7.1	Choice of raw materials and supplies	92
7.2	The manufacturing process	92
7.3	Use of products	92
	References	93

CHAPTER 3

Outi Kimari

Tissue

The term "tissue" describes products made from lightweight, dry creped and some non-creped paper such as toilet paper, kitchen towels, handkerchieves, facials, napkins, hand towels, wipes. Products of such a kind derive from a one-ply, semi-finished, wet-laid tissue base paper, that is predominantly composed of natural fibers. The origin of the fibers may be virgin or recycled. Properties of the tissue base paper give to its resulting products the typical high tensile energy absorption together with a good textile like flexibility, surface softness, comparatively low bulk density, and high ability to absorb liquids. Nonwovens do not belong to the group of tissue papers.

Paper has been used for hygiene purposes for centuries. In the case of tissue papers, this is still the most important use today. However, tissue paper as we know it today was not manufactured in the United States until after mid–1940s. In Western Europe, large-scale industrial manufacture of tissue products began in the beginning of 1960s, with Finland as one of the forerunners.

1 Tissue markets

In 1996, global consumption of tissue totaled roughly 17 million t. The consumer sector accounts for about 70% of the total and the away-from-home sector for the remaining 30%. The tissue market in 1996 was worth about US$ 40 billion.

Over the past twenty years, global consumption of tissue has grown at an average of 4% a year. Growth has been fastest in the countries of Asia (Fig. 1).

Calculated per capita, tissue consumption worldwide is roughly 3 kg a year. Consumption is highest in the United States, where it is over 20 kg a year. In Africa, annual consumption per capita is only 0.36 kg. In the Nordic countries, the figure is around 15 kg (Fig. 2).

There are currently more than 700 mills producing tissue worldwide. The number of tissue machines in production is around 1 400. These have a total annual capacity of over 19 million t. The ten largest manufacturers produce over half of all the world's tissue.

Global tissue manufacturing capacity is forecast to rise by almost 2.6 million t (Fig. 3) by the year 2010 – about one quarter of the anticipated growth in demand. However, capacity and demand will not increase at the same rate everywhere, and Western Europe, for example, could experience over-capacity.

CHAPTER 3

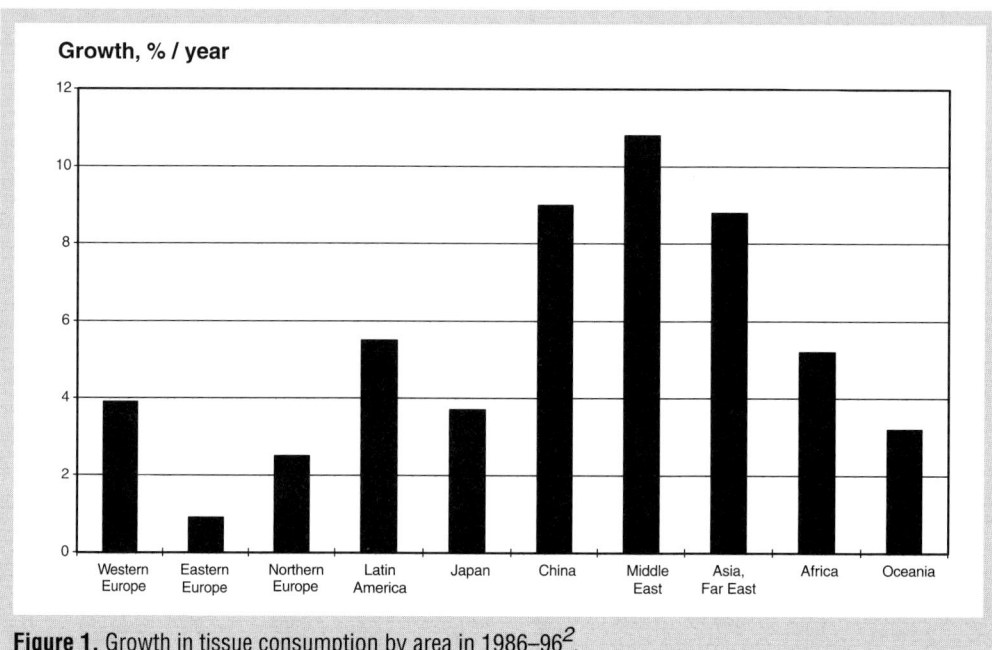

Figure 1. Growth in tissue consumption by area in 1986–96[2].

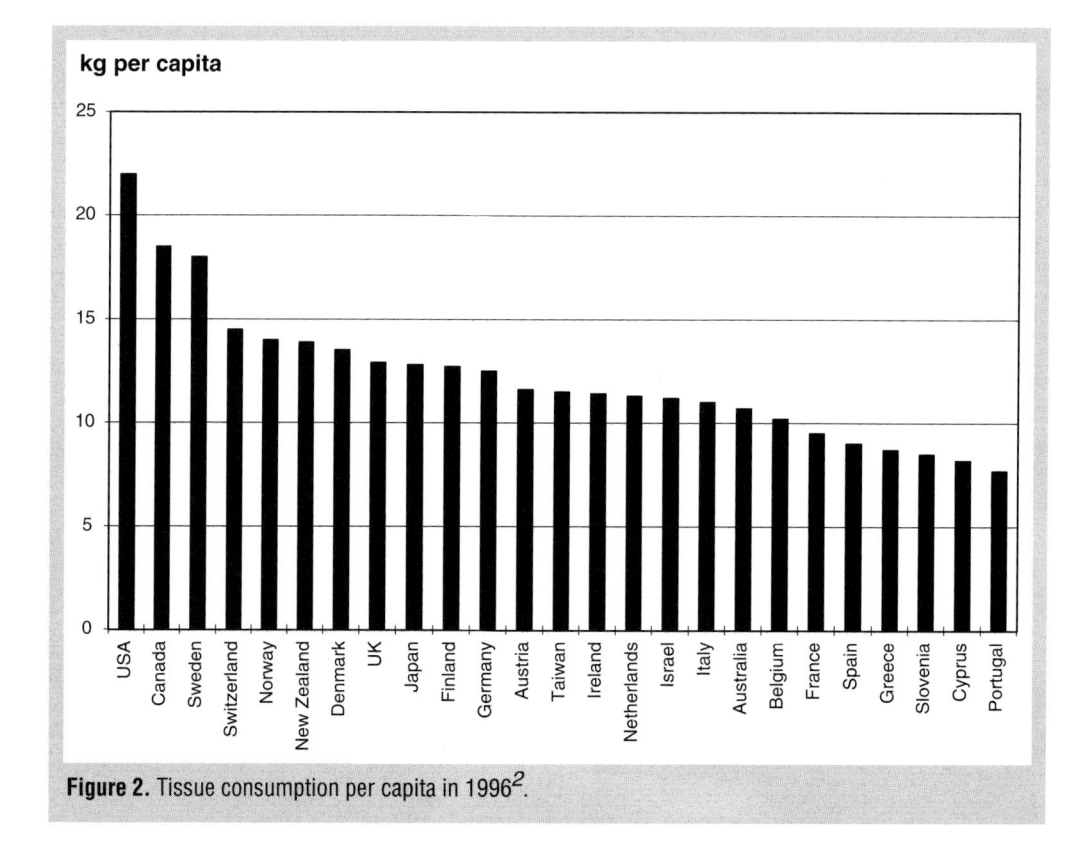

Figure 2. Tissue consumption per capita in 1996[2].

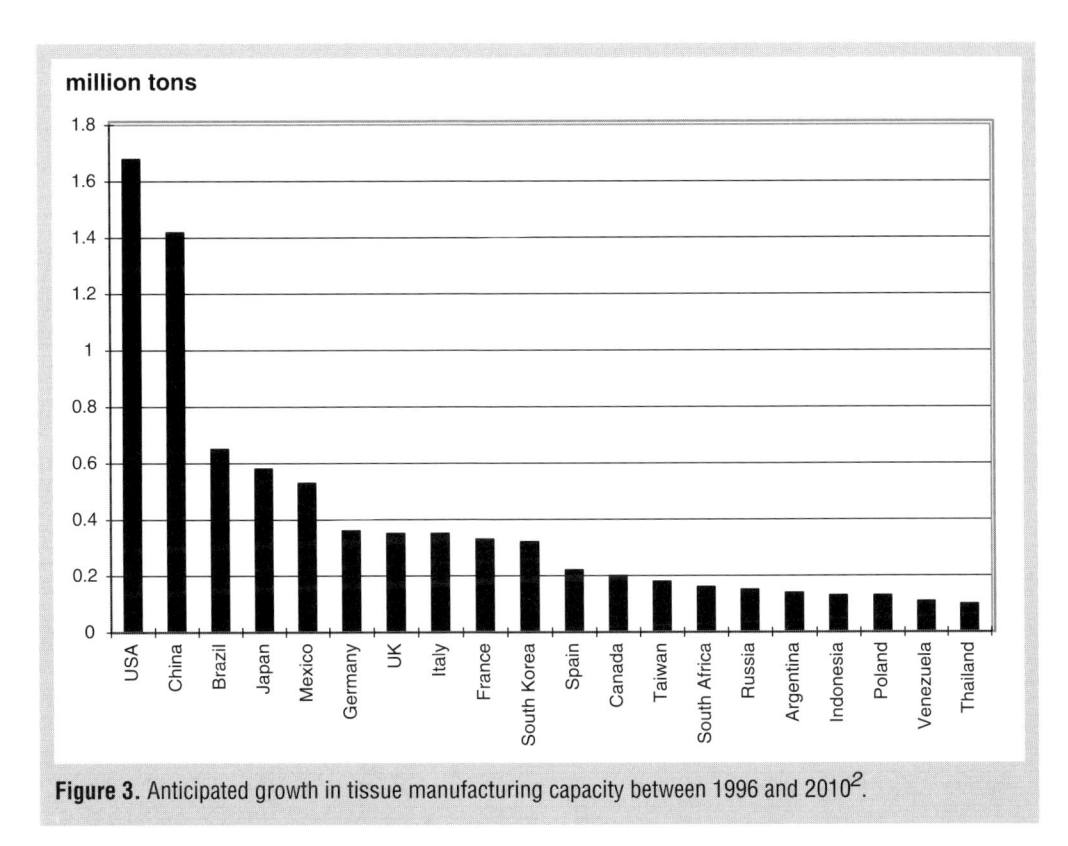

Figure 3. Anticipated growth in tissue manufacturing capacity between 1996 and 2010[2].

2 Raw materials for tissue production

As in the production of other grades of paper, tissue is made from stock in which fibers are mixed with water. Agents that will improve the paper's characteristics can be added to this stock. The most common of these are wet strength size to give the paper greater wet strength, pigments to make the paper a certain color, and tinting dyes and optical brighteners to give the desired shade. The addition of anti-foaming agents and chemicals for pH control make the paper easier to run on the paper machine. All the agents employed in the production of tissue are approved for use in conjunction with food.

2.1 Pulps

The pulp used in the production of tissue is made from either primary fiber or recycled fiber pulp.

2.1.1 Virgin fiber pulp

Virgin fiber pulp produced chemically by either sulfate or the sulfite process is suitable for making tissue. The type most widely used worldwide is sulfate (kraft) pulp, and in Finland all tissue is made from this pulp.

CHAPTER 3

Chemical pulp may be either bleached or unbleached. Finnish tissue mills exclusively use bleached chemical pulp. Bleaching improves the product's brightness and absorption properties, gives it longer life, and reduces the level of impurities. Today, all environmentally conscious tissue mills use pulp that is either bleached with oxygen chemicals, i.e., totally chlorine-free (TCF) pulp, or elemental chlorine-free (ECF) pulp.

The chemical pulp used for tissue production is made from either softwood or hardwood fiber. Most softwood pulp is pine, although it may contain some spruce. Hardwood pulp can be made from woods such as birch, eucalyptus, and beech. Finnish mills normally use birch. Softwood pulp has longer fibers and gives the paper strength, whereas hardwood fibers impart softness.

Chemithermomechanical pulp (CTMP) can also be added to the papermaking stock to make the paper more absorbent and bulky.

2.1.2 Recycled fiber pulp

There are two main grades of recycled fiber pulp: selected, i.e., containing mainly chemical pulp fiber (office waste), and ordinary (mixed waste), i.e., made from newspapers and magazines. Both are used in tissue manufacture.

Tissue was one of the first paper grades to be made from recycled fiber pulp. Finland's first recycled fiber pulp mill started production in the 1970s.

Selected recycled fiber pulp (based on chemical pulp fiber) is made from recovered paper of better quality, such as office waste. The resulting pulp is light in color, having a brightness of at least 80%.

Ordinary recycled fiber pulp is made from collected household recovered paper, which consists largely of newspapers and magazines. It usually has a brightness of 60%.

2.1.3 Pulp mix

The type of chemical pulp fibers or recycled fibers used depends on the quality requirements placed on the base tissue. For example, base paper intended for facial tissue contains more hardwood fiber for greater softness, while that for paper towels and industrial wipes contains more softwood fiber for greater strength. Higher-quality recycled fiber products are made from selected recycled fiber, while lower grade products contain recycled fiber pulp made from household waste paper.

Correspondingly, the proportions of primary and recycled fiber used also depend on the quality requirements. The extremes are tissue made from 100% primary fiber and tissue made from 100% recycled fiber. The highest quality products are made from pure chemical pulp made from primary fiber. The resulting paper is strong, absorbent, and soft.

3 Uses of tissue

3.1 Hygiene products

The biggest and most important use for tissue is in hygiene products:

- Bathroom tissue
- Kitchen towels
- Folded paper towels
- Industrial wipes
- Paper handkerchiefs
- Table napkins
- Facial tissue
- Disposable diapers
- Incontinence and hospital pads
- Sanitary napkins
- Refreshing towelettes.

These products are manufactured for both the consumer and the away-from-home markets. In 1996, about 70% of all tissue produced worldwide was sold on the consumer market. However, the away-from-home market is growing rapidly.

In the hygiene sector, tissue intended for wiping and drying competes with fabric wipes, cloths, cotton wads, and automatic hot-air dryers.

3.1.1 Bathroom tissue

Bathroom tissue is the biggest single product made from tissue in both the consumer and away-from-home sectors. In Western Europe, for example, bathroom tissue accounts for over 50% of tissue consumption. In North America, the figure is over 80%.

The base tissue used to make bathroom tissue has a basis weight of 14–22 g/m^2. Bathroom tissue is made with one, two, three, or four plies. The width of a roll (sheet width) is 100–115 mm, and the length of the sheet is 90–150 mm. Rolls intended for the consumer market contain 180–280 sheets. Those for the away-from-home market can have up to 2 000 sheets per roll, although the number is usually 600–800.

The composition of bathroom tissue varies from 100% primary fiber (i.e., chemical pulp) to 100% recycled fiber. The base paper used to make this type of tissue has to disintegrate in the sewage system and, therefore, is not strengthened by the addition of wet-strength sizes.

Bathroom tissue may be smooth or embossed, unprinted or patterned, pure white, off-white, or tinted. There are many variations in raw material composition, color, and embossing, with the choice depending very much on the country and market in question.

CHAPTER 3

3.1.2 Kitchen towels

Kitchen towels are the second biggest product for the consumer sector. Consumption varies greatly from one country to another, but is well below that of bathroom tissue.

Kitchen towels usually have a basis weight of 20–24 g/m^2. Sheets are 220–280 mm wide and 250–280 mm long. The number of sheets to a roll is normally between 54 and 100.

Most kitchen towels are two-ply. More than two plies is rare, but some one-ply towels are made. As with bathroom tissue, the base tissue from which kitchen towels are made, can, in theory, vary from 100% chemical pulp fiber to 100% recycled fiber and includes everything in between. In view of the strength requirements, the pulp from which kitchen towels are made normally contains at least some primary fiber, i.e., chemical pulp. Wet-strength resin is added to further improve the paper's strength.

Kitchen towels can be either printed or unprinted. It is important for kitchen towels to have high absorbency, and for this reason they are usually embossed.

3.1.3 Paper towels and industrial wipes

Paper towels and industrial wipes are primarily intended for the away-from-home market. They have a huge range of different uses, ranging from wiping grease in engineering workshops to personal hygiene in hospital surgical wards.

As with bathroom tissue, the fiber composition of paper towels and industrial wipes can range from 100% pure chemical pulp to 100% recycled fiber. For most purposes, resin is added to improve wet strength.

Paper towels normally have either one or two plies. The tissue for one-ply towels has a basis weight of 33–50 g/m^2 and that for two-ply towels has a basis weight of 22–24 g/m^2. The surface may be either smooth or embossed, printed or unprinted.

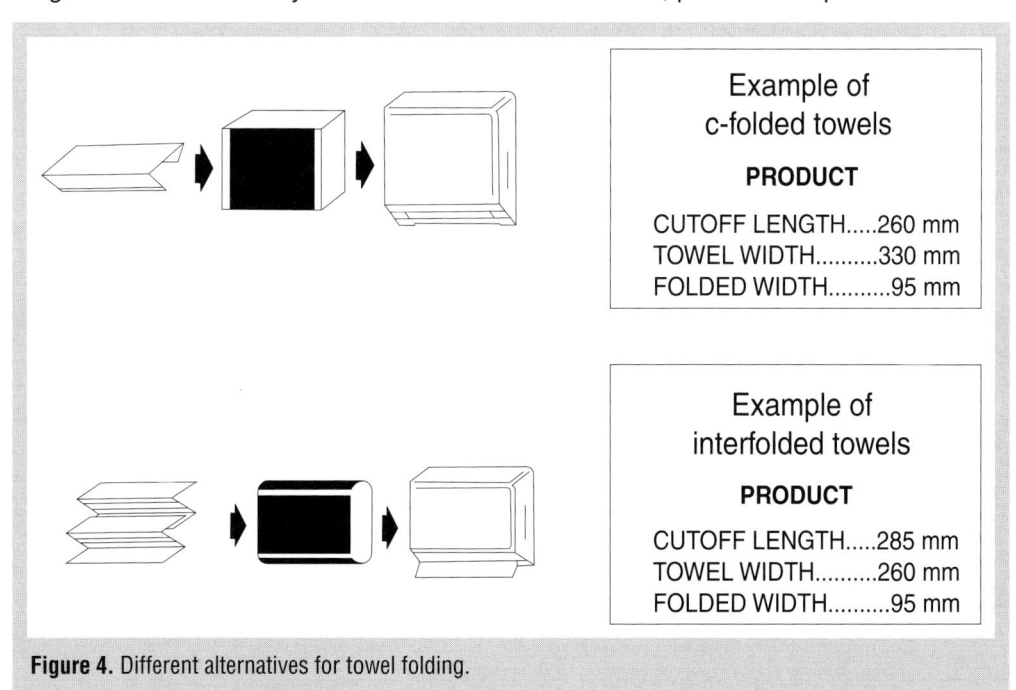

Example of c-folded towels

PRODUCT

CUTOFF LENGTH.....260 mm
TOWEL WIDTH..........330 mm
FOLDED WIDTH..........95 mm

Example of interfolded towels

PRODUCT

CUTOFF LENGTH.....285 mm
TOWEL WIDTH..........260 mm
FOLDED WIDTH..........95 mm

Figure 4. Different alternatives for towel folding.

Paper towels can be folded in different ways, the most common being C-fold and interfold (Fig. 4). Unfolded paper towels are also produced. The size of the sheet varies according to the type of fold. The correct dispenser or holder plays an important part in the use of paper towels.

The type of fold affects the amount of paper needed for a single sheet, and therefore the price of the product. C-folded paper towels, which are common in the Nordic countries, require more base tissue than inter-folded paper towels. The fold also determines how convenient the towel is to use. An inter-folded towel, for example, is more convenient than a C-folded version because it opens as it is taken from the dispenser and brings the next towel down in the ready position. C-folded towels, on the other hand, are merely stacked in the holder, from where they emerge unopened.

Industrial wipes are normally in the form of rolls. They consist of one, two, or four plies and are made from 25–50 g/m^2 base tissue. The paper can have a smooth or an embossed surface.

The width of a roll varies between 200 and 500 mm. The largest rolls can have diameters of over one meter. Rolls can also be supplied coreless, in which case the wipes are pulled out from the center.

3.1.4 Facial tissue and handkerchiefs

Facial tissue and handkerchiefs are made from the lowest basis weight tissue (14–18 g/m^2). The surface is often made smoother by light calendering. Facial tissue is usually two-ply while handkerchiefs consist of two or three plies. Because of the high quality requirements, the base tissue for most facial tissues and handkerchiefs is made either entirely from pure chemical pulp or from a mixture of pure chemical pulp and selected recycled fiber.

Facial tissues are normally supplied in a box that also acts as a dispenser. The box is either square or rectangular. The latter can also be used for larger sizes (known as man-size).

Paper handkerchiefs are folded either square (the traditional way) or made pocket size. For square handkerchiefs, the sheet size ranges from 25 cm x 25 cm to 29 cm x 29 cm.

Paper handkerchiefs and facial tissue are used in different ways in different countries. Paper handkerchief consumption is high in countries like Germany and Finland, whereas facial tissue is a high-volume product in the United States and Britain. Facial tissue is used not just by ordinary consumers but also in places like hotels.

3.1.5 Table napkins

Table napkins are produced with one, two, three, or four plies. The size and type of fold vary greatly – from small coffee cup napkins to large table napkins for dinner parties.

Table napkins may be chosen to match the decor, and thus come in many colors and patterns. Tissue intended for table napkins can be dyed in bold colors to match the prevailing fashions. As with other tissue products, the composition of the raw material can vary tremendously. The highest quality napkins have a high content of chemical pulp together with some selected recycled fiber.

CHAPTER 3

3.2 Other tissue products

Base tissue is also used to make coffee filter papers, cigarette filters, and padding for meat packaging. Tissue can be laminated with nonwoven fabrics to make material for disposable clothing for use in health care, etc.

3.3 Quality requirements

Tissue is used to make a huge range of products, and the quality requirements placed on the base paper and the converted products vary according to the purpose of the product and the expectations of the user. In some products, softness can be the key property, while in others it might be strength. In some situations, the product might have to absorb water, in others perhaps oil. Base tissue also has to meet certain quality demands for it to be suitable for converting.

The main quality requirements, which can be measured from the tissue itself, are:

- Basis weight
- Absorbency
- Softness
- Thickness (bulk)
- Tensile strength
- Brightness
- Stretch
- Appearance (stylishness).

3.3.1 Basis weight

Depending on the end use for the product, base tissue can have a basis weight from 12 to 50 g/m^2. Low basis weights are used in multi-ply products where softness is important (facial tissue, top-quality bathroom tissue). Single-ply industrial wipes, which have to be strong, are made from tissue with a higher basis weight. With several products, the aim is to achieve more performance with less material.

3.3.2 Absorbency

Absorbency is normally measured using water. However, measuring methods and points of view vary with the country and the product in question.

The properties normally measured are absorption capacity and absorption rate, which are important in products such as towels and wipes. Absorption capacity reflects how much water the tissue can absorb (g water/g paper). Absorption rate measures how quickly the product can absorb water. The units are seconds per cm (s/cm). In certain products, the liquid used for the measurements might be oil.

Absorbency is also affected by the papermaking process, the paper's raw material composition, and the paper's bulk. A one-ply tissue made from recycled fiber and embossed in the normal way has an absorption capacity of about 4 g/g; that of the top-quality kitchen wipes might be as high as 18 g/g.

3.3.3 Softness

Softness is very much a subjective property for which there is no generally accepted method of measurement. Each mill tends to use its own methods, which are based on comparisons. One such method is Hand Feel (HF), in which softness is assessed on a scale from 2 to 6. High-quality bathroom tissue has an HF score of 6, while ordinary tissue with a high recycled fiber content has a score of 2.

The HF score basically measures the surface softness, which is related to smoothness. How soft a product feels also depends on its thickness (bulk) and its flexibility.

Softness is very important for products such as bathroom tissue and facial tissue.

3.3.4 Thickness and bulk

Compaction of the fiber network and inter-fiber bonding are undesirable in the base tissue structure. The aim is to produce a sheet with the highest possible specific volume, i.e., bulk. For a given product, web thickness is the same as bulk, which is the word used in the industry to refer to thickness. Thickness is usually measured by compressing a sample consisting of several plies under a small load between two measuring heads.

For converted tissue products, the target bulk is usually set in terms of the desired roll diameter – a certain number of sheets should give a certain roll diameter. The target bulk set for a converted tissue product can be influenced by embossing. In the case of unembossed products, the bulk of the base paper must be correct if the targets set for roll diameter and sheet number are to be achieved.

3.3.5 Tensile strength

Tensile strength is measured in both the machine and cross-machine directions of the web, both wet and dry. The industrial wipes used in places like engineering workshops and car repair shops need to have a high tensile strength to stand up to heavy-duty wiping.

On a traditional tissue machine, the three properties softness, bulk, and tensile strength can never all be optimized. As tensile strength increases, softness and bulk decrease, and vice versa. This means that for a particular product, the targets set for these properties are always something of a compromise, the priority given to each depending on the intended use for the product.

3.3.6 Brightness, whiteness

The brightness of base tissue varies from 50% to 88%, as measured according to the ISO standard. Products made from bleached chemical pulp have the highest brightness. In the case of 100% recycled fiber, brightness is about 50%.

Often more important than ISO brightness is the product's whiteness, which is a measure of how pale and of what shade the paper is perceived by the human eye. To improve the whiteness value, tinted dyes and optical brighteners can be added to the papermaking furnish.

3.3.7 Crepe ratio, stretch

The crepe ratio reflects how much a paper has shortened during creping. (See also "Tissue production process", "Drying section, creping".) The figure is usually between 10 and 30%. Creping is used to adjust the paper's stretch and thickness, both of which have a marked effect on softness and absorbency. Stretch is also important in determining the suitability of base tissue for converting.

3.3.8 Appearance

The appearance, or stylishness, of converted tissue products is another important visual feature. This involves regularly monitoring and measuring factors such as the number of sheets, roll diameter, the cut of the roll edges, perforation, embossing, color, and the pattern printed on the paper. The appearance and quality of the pack are another part of the quality image.

3.3.9 Quality requirements for converting

Aside from the quality requirements discussed above, base tissue must meet certain other quality requirements if it is to be efficiently converted. In many cases, these are similar to the requirements imposed by the product's use.

To begin with, all basic dimensions such as basis weight and bulk must be correct if the end product is to meet the quality demands set for it. To some extent, deficiencies can be overcome during converting processes, but this makes converting less efficient.

The paper has to be sufficiently strong and the formation even, i.e., the web must be free from holes, if it is to run smoothly on converting machines. Machine rolls for converting must be creped over the correct area so that the paper cuts properly for products such as paper towels. The fewer breaks there are in the base paper roll, the more efficient is the converting process. Other important factors include the right moisture content and low linting (dust formation).

4 Special features of the tissue production process

The tissue-making process is basically the same as for other grades of paper. The tissue paper machine consists of a wet end (headbox, forming section), a press section, and a drying section[4]. The fiber slurry, which at this stage is 99.8% water, is sprayed from the headbox onto the wire of the forming section. The fibers gradually form a web as they pass on through the press

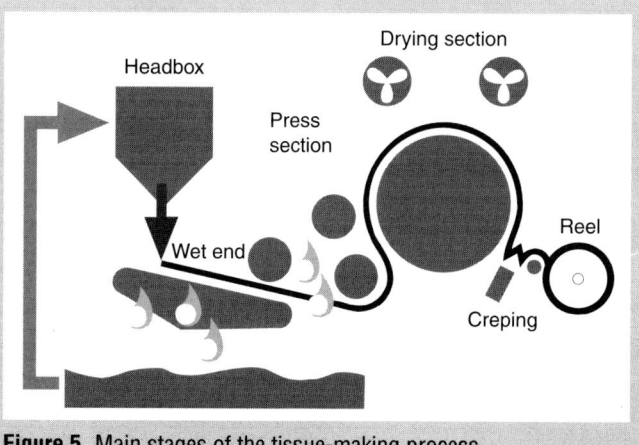

Figure 5. Main stages of the tissue-making process.

and drying sections. Most of the water is removed along the way, so that the final tissue web contains only 4%–8% water.

A tissue paper machine is much shorter than an "ordinary" paper machine. A typical feature of tissue paper is the creping that takes place after the drying section (Fig. 5). Other features special to the tissue-making process can be found in the forming and drying sections: the forming section is short, while the drying section often consists of a single Yankee cylinder.

Despite technological advances, the widths of tissue machines have not increased significantly over the years. Tissue machines are much narrower than fine paper machines, for example. The idea has been to keep the width of the tissue web in line with that required by converting lines. The most common machine widths are 280 cm, 330 cm, and 500 cm. A few machines have widths of around 750 cm.

Tissue machines normally operate at between 800 and 2 000 m/min. Crescent former type machines have an average speed of 1 500–1 700 m/min. Speed is determined not just by the type of machine (principally its forming section) but also by the paper being produced. Higher basis weights are generally produced more slowly than lower ones because drying capacity is the main bottleneck.

In principle, a given tissue machine can be used to produce papers for a wide range of different uses. However, in practice, tissue machines normally specialize in particular grades, for example, certain basis weights. This makes it easy to adjust the machine to meet the special demands of the papers being produced. For example, multilayer machines are particularly suitable for producing papers that need to be extremely soft. (See also "Multilayer headbox".) Machines with after-drying sections, on the other hand, are used to make thicker papers that have to be especially strong. (See also "Drying section, creping".)

4.1 Multilayer headbox

If necessary, the stock in the headbox of a tissue machine can be divided into two or three layers (Fig. 6). This allows the characteristics of the paper to be adjusted. For example, one layer can be produced from long-fiber pulp to add strength, while the other layer can be short-fiber pulp for softness. When the web leaves the drying section, it can be reeled so that the soft side is uppermost and the strong side underneath, allowing the softness-strength requirement to be met in the optimum way.

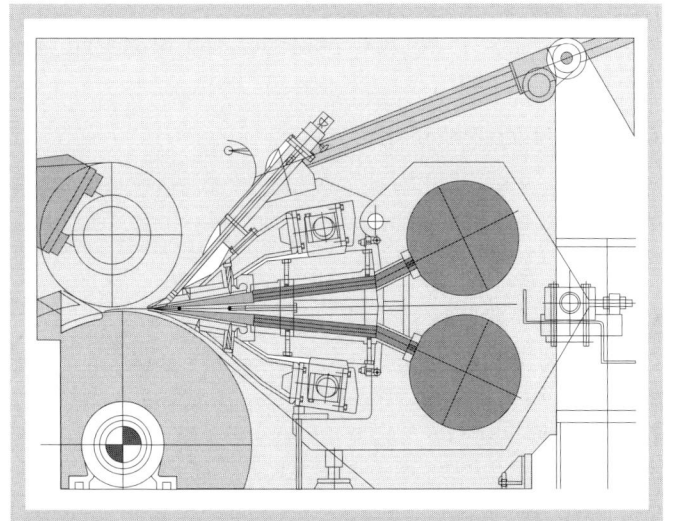

Figure 6. In a multilayer headbox, the stock can be divided into two or three layers, allowing the characteristics of the tissue to be controlled[6].

CHAPTER 3

4.2 Tissue forming sections

In terms of forming section, there are four types of tissue machine:
- Fourdrinier machines
- Suction breast roll machines
- Twin-wire machines
- Crescent former machines.

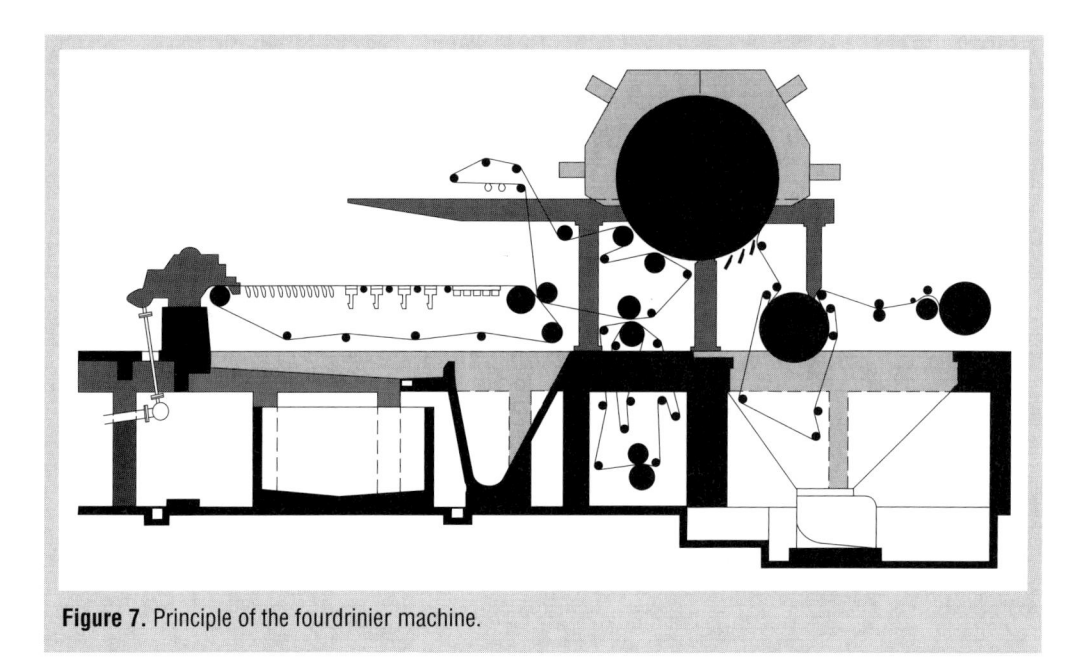

Figure 7. Principle of the fourdrinier machine.

4.2.1 Fourdrinier machines

The oldest tissue machines are fourdrinier machines, on which water is removed from the web simply by gravity and suction drainage (Fig. 7). There are still a few of these machines running today.

4.2.2 Suction breast roll machines

In a suction breast roll machine, water is forced out of the web in the space between the top of the headbox and the wire (Fig. 8). Water removal is controlled by suction. This type of machine was dominant up until the 1970s.

Tissue

Figure 8. Principle of the suction breast roll machine.

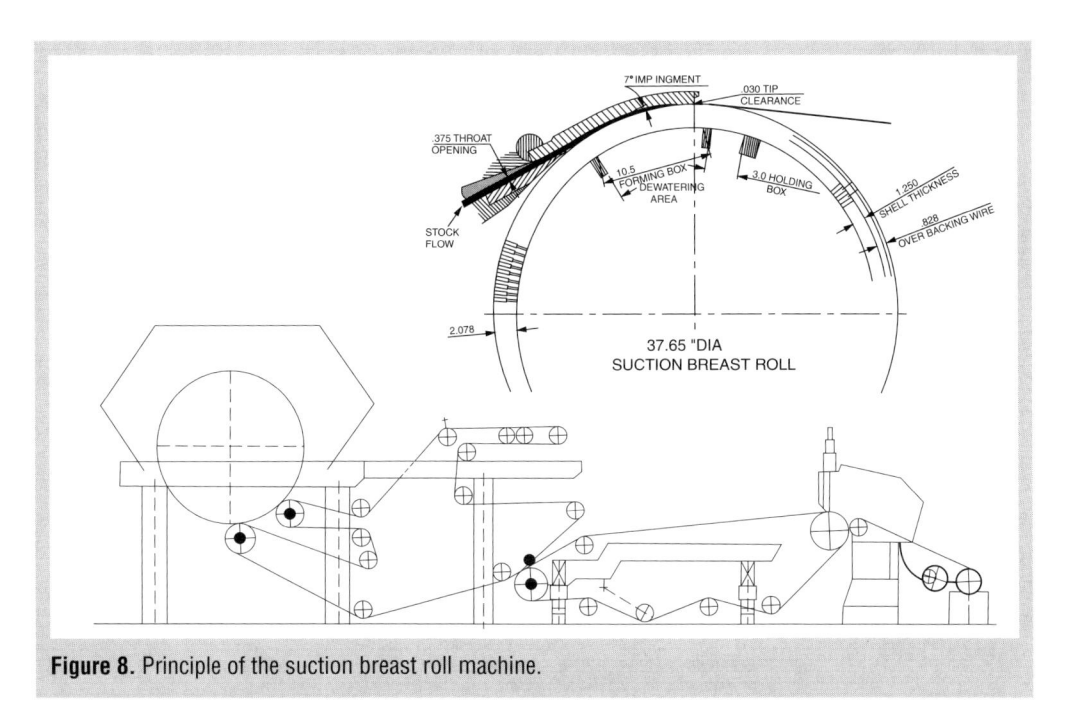

4.2.3 Twin-wire machines

The twin-wire machine, in which the stock is sprayed from the headbox in between two wires, provides more effective dewatering. Many of the tissue machines currently in operation are twin-wire machines. Although there are differences between different manufacturers' machines, the basic principle is the same (Fig. 9).

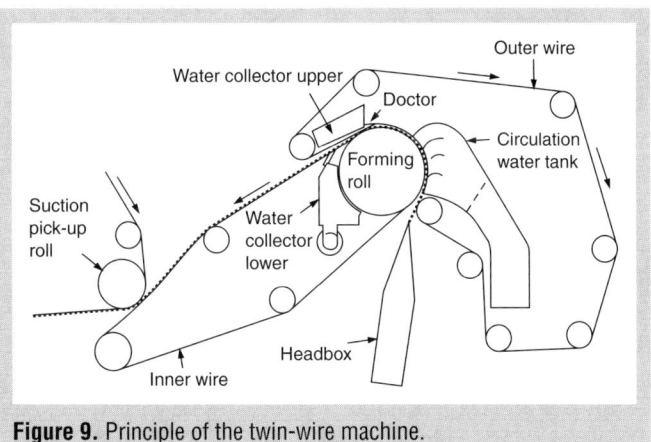

Figure 9. Principle of the twin-wire machine.

4.2.4 Crescent former machines

The crescent former (Fig. 10) was originally developed in the 1960s, but was protected by patent until the late 1980s, after which it rapidly spread throughout the tissue industry[3]. In this arrangement, the stock is sprayed between the wire and a felt. The result is superior runnability because the web no longer has to be taken off the wire and placed onto the felt.

CHAPTER 3

Figure 10. Principle of the crescent former.

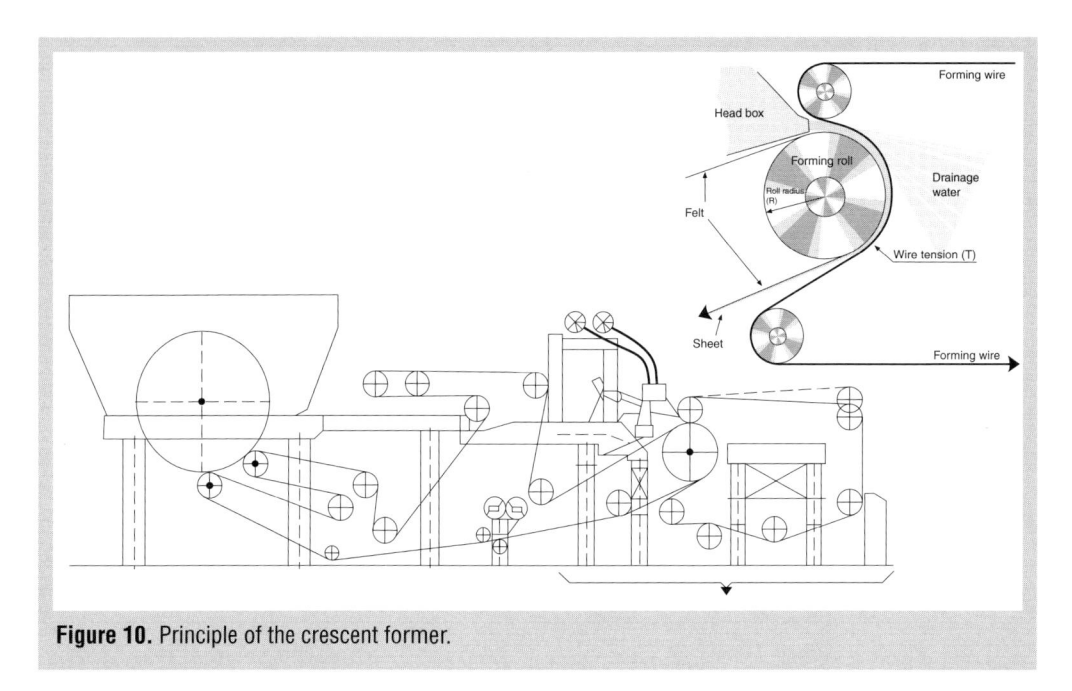

4.3 Drying section, creping

On a tissue machine, the drying section comprises a single steam-heated Yankee cylinder fitted with a hot air hood. The cylinder can have a diameter of up to 5.5 meters and the pressure inside is around 8 bar. Air is blown onto the web at temperatures up to 550°C and speeds up to 170 m/s.

After the yankee, there can also be what is termed an "after-drying" section. However, in most cases, the yankee dryer is sufficient.

Creping, which is performed by the yankee's doctor blades, is a critical stage in determining the characteristics of the final paper. Creping involves spraying the cylinder with a suitable amount of adhesive and pressing it against the surface of the yankee cylinder. When dry, the web can be removed from the cylinder surface by doctoring. The crinckle, or crepe, remains in the paper[1].

Figure 11. Tissue is creped by the doctor blades of the yankee cylinder. Creping involves spraying the cylinder with a suitable amount of adhesive and pressing it against the surface of the yankee cylinder. When dry, the web can be removed from the cylinder surface by doctoring. The crinckle, or crepe, remains in the paper[1].

amount of adhesive and pressing the paper web against the surface of the yankee cylinder (transfer from the felt). When dry, the web can be removed from the cylinder surface by doctoring. The crinkle, or crepe, remains in the paper. Creping depends on factors such as the strength of the adhesive, the difference in speed between the yankee cylinder and the final section of the paper machine, doctor blade geometry (shape and position), and the fiber raw materials used in the stock.

Creping increases the paper's thickness (bulk) and absorbency (Fig. 11). Coarse creping produces papers suitable for wiping oil and grease, while finely creped paper is soft and is used for products such as facial tissue.

4.4 Through-air drying system (TAD)

Tissue machines employing a through-air drying system (Fig. 12) form their own group. They feature a perforated through-air drying cylinder positioned after the forming section but before the yankee cylinder. Air is blown through onto the tissue web as it passes round the cylinder. There is no wet pressing stage. The paper web is made to adhere to the yankee cylinder almost dry. Creping then takes place in the normal way.

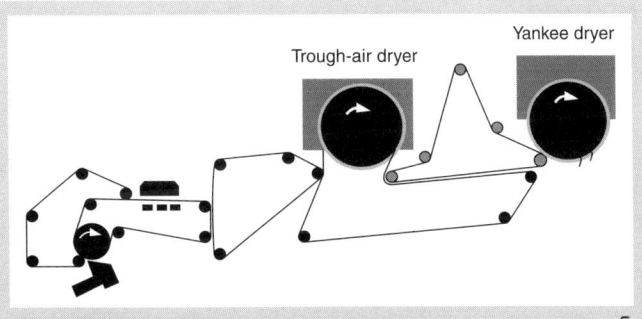

Figure 12. Tissue machine employing a through-air drying system[5].

A method has recently been developed in which traditional cylinder creping is replaced by speed differences between machine fabrics.

Through-air treatment greatly improves the paper's softness and doubles its absorption capacity. Made in this way, products such as bathroom tissue and kitchen towels are softer, bulkier, and more absorbent even as single-ply than multi-ply products produced by the traditional process.

Through-air drying machines have been in use longest in the United States. In Europe, they are only just being introduced on a wide scale. This might be because of the high capital cost of these machines and also because the through-air drying system raises energy costs.

5 Tissue finishing

The most common finishing unit on a tissue machine is re-winding. This involves winding the machine roll again in two or more layers. The same operation can be performed at the start of a converting line.

The winder also cuts the machine roll into smaller rolls, which are more convenient for converting purposes. The widths depend on the width of the converting machine. Converting lines producing bathroom tissue and kitchen towels normally have widths of 170, 260, or 340 cm. At some tissue mills, the paper machine and converting machine have the same width, and no winder is therefore needed.

6 Tissue converting

Several stages are needed before the re-wound and cut-to-size rolls from the paper machine finishing section are made into a range of different hygiene products. This process is referred to as converting. The main stages (as shown in Fig. 13) are:

- Embossing
- Printing
- Perforation
- Winding and tail sealing, folding
- Log sawing
- Packaging.

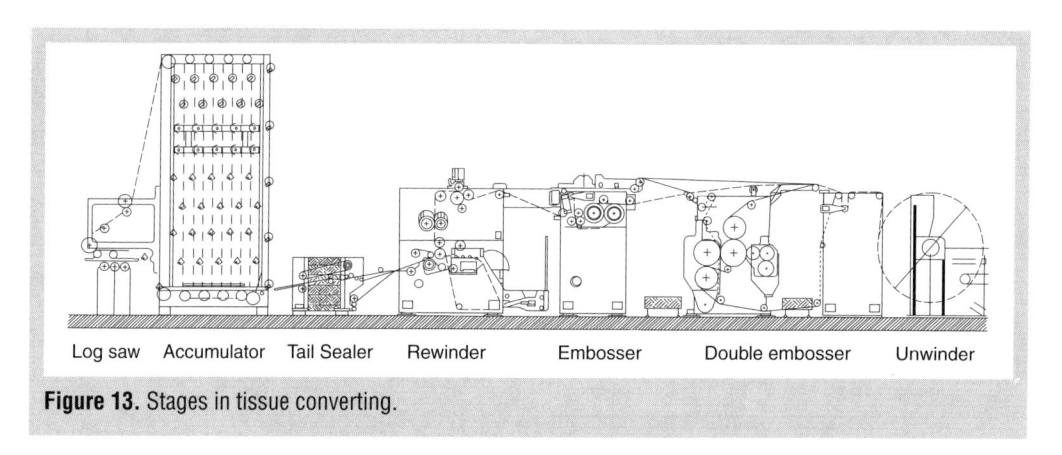

Figure 13. Stages in tissue converting.

6.1 Embossing

Most tissue products – bathroom tissue, kitchen towels, paper towels, and industrial wipes – are embossed, which means the plies are pressed together either completely or just at the edges. This gives the product greater softness and absorbency. Products can sometimes be embossed for decorative reasons.

There are three main types of embossing:

- Traditional embossing
- Nested embossing
- Foot-to-foot (also known as tip-to-tip) embossing.

6.1.1 Traditional embossing

In traditional embossing, all tissue plies are embossed simultaneously. The two embossing rolls can be both of steel or one might be steel and the other rubber. This was virtually the only embossing method used up until the early 1980s, and is still the most common way of embossing bathroom tissue.

6.1.2 Nested embossing

In nested embossing, each tissue ply is embossed separately, after which the plies are bonded together. This results in bulk that is almost double that achieved with traditional embossing. Nested embossing was introduced in the early 1980s and is widely used for kitchen towels.

6.1.3 Foot-to-foot embossing

Foot-to-foot embossing (also known as tip-to-tip and point-to-point embossing) is a more advanced form of nested embossing. Here again, each ply is embossed separately and the plies then bonded together. The bulk achieved is 2.5 times greater than with traditional embossing. The positioning of the highest points of the surface pattern on the embossing rolls opposite each other means suction pockets are formed between the plies, and this improves the product's rate of absorption.

6.2 Printing

Printing takes place either on the converting line or on a separate printing press. Printing can take the form of various patterns or designs (bathroom tissue, kitchen towels, table napkins) or names (table napkins, paper towels). Most converting lines are capable of printing up to four colors, although printing is normally restricted to either one or two colors. Table napkins can be printed in as many as 6–8 colors. The pattern or design can cover the entire product or just a part of it, for example, the edge of a roll or sheet.

6.3 Perforation

Products such as bathroom tissue and kitchen towels that are supplied in rolls are perforated to make the sheets easier to separate. This is done on the converting line by perforation knives. On modern converting lines, the interval between perforations (i.e., the sheet length) is easily controlled.

6.4 Winding and tail sealing or folding

Winding gives the rolls of tissue products (bathroom tissue, kitchen towels, industrial wipes) their final shape. Here, rolls are produced with the desired amount of paper – usually determined by the number of sheets or the number of meters – by winding the paper onto a paper core and sealing the tail. Sheet products such as paper towels, facial tissue, and handkerchiefs are folded to give the desired shape.

6.5 Log sawing

Before packaging, the products are cut to the desired width. The final width of rolls is 10–60 cm.

6.6 Packaging

The final step in the converting chain is to package the rolls and stacks of sheets into consumer and transport packs. The packs are wrapped in either plastic or paper, and for transport are placed in plastic sacks or corrugated board boxes.

CHAPTER 3

7 Environmental impact of tissue manufacture

The manufacture and converting of tissue affect the environment in several ways at each stage of the process. Environmental issues are encountered in:

- Choice of raw materials and supplies
- The manufacturing process
- Use of products.

Environmental labels (national, Nordic Swan, EU flower) are effective in encouraging the consideration of environmental issues throughout the production chain and in reducing the environmental impact of products over their entire life cycles.

7.1 Choice of raw materials and supplies

The basic raw material for tissue production is wood. Wood is needed to produce chemical pulp, and it is also the raw material from which recycled fiber is originally made.

In wood procurement, the environment is taken into account by ensuring that felling is in line with forest growth and that forest management in general is based on sustainable development.

In the production of chemical pulp and recycled fiber pulp, emissions to the environment are minimized by recycling process waters and a whole range of other measures. Chemical pulp is bleached without chlorine. And deinking waste is utilized as far as possible.

Tissue manufacture and converting also require packaging materials, printing inks, and so on. The materials chosen are suitable for recycling, burning, or composting.

7.2 The manufacturing process

The tissue-making process affects the environment through process emissions to both air and water. Discharges to waterways are minimized by recycling process waters, treating wastewater biologically, and seeking to reduce the amount of water used in the process.

Emissions to air arise from the generation of heat and electricity needed in the tissue-making process. To reduce the impact, environment-friendly fuels are chosen and various filters are used to remove impurities from the flue gases.

7.3 Use of products

Users of tissue products take the environment into account by choosing the most suitable product for each purpose. This cuts paper consumption and reduces waste volumes.

Used tissue is biodegradable and can therefore be composted and utilized.

References

1. Gavelin, G., Creped tissue, Swedish Forest Industry Association, Manual Y-306, 1992.

2. Uutela, E., and Bognar, R., "Tissue Market Developments," 1998 Tissue Making, Karlstad, Sweden.

3. Hender, B., "Microtex Forming Fabrics^R Increase Productivity of Crescent Formers," 1998 Tissue Making, Karlstad, Sweden.

4. Klerelid, I., "SymFlo™ TIS Headbox, Design and Running," 1998 Tissue Making, Karlstad, Sweden.

5. Jewitt, D., "Machine Configurations for Through Air Drying," 1998 Tissue Making, Karlstad, Sweden.

6. Greimel, R. and Tietz, M., Twogether **5**:61 (1998).

    Ryöti, L., Karppinen, J., Virkajärvi, J., and Riekkola, H., Personal communications.

CHAPTER 4

Air-laid paper

1	Introduction	95
2	Process	96
3	Product characteristics and applications	97

CHAPTER 4

Tapio Niemi

Air-laid paper

1 Introduction

When talking about paper, people normally think of the papers of everyday life, i.e., newspapers, magazines, books, writing and copying papers, and paper and board for packaging. Tissue papers are normally understood as toilet papers, kitchen towels, and other wiping papers. All these papers are generally called "wet-laid" papers, i.e., the sheet forming process is based on forming the fibrous sheet structure in water. The term "air-laid" paper does not tell too much to the majority of people. The Russian scientists Dimitriev and Bondarenko can be called pioneers of the air-laid technology. They introduced the theoretical principles of air-laid papermaking in 1931.

The word "nonwoven" means basically textile-like material made of fibers with technology which binds fibers together without weaving. The fibers can be natural, man-made or a mixture of both and they are carried to the web forming section by means of water or air.

Originally nonwovens were developed to substitute textiles in some less demanding applications. Today nonwovens industry is one of the fastest growing industries in the world while textile industry is declining.

In spite of textiles nonwoven fabrics have substituted paper, especially tissue paper in some applications. Long fibers in nonwovens serve very well in reinforcing resulting in higher strength properties compared with paper. Wiping materials are a good example of those applications.

Air-laid paper falls in between paper and nonwoven. From the fiber origins (wood pulp) point of view it is paper but from end use point of view it belongs to nonwovens. Manufacturing process resembles both paper and nonwovens. As an industry air-laid is generally referred to as the nonwovens industry.

The beginning of active working air-laid technology was in the 1960s and 1970s with the development of the three main air-laid processes currently used: K. Kroyer, Denmark; Dan-Webforming, Int. (Scan-Web), Denmark; and Honshu Paper, Japan.

Typically, the beginning of air-laid production in the 1970s experienced great technical difficulties, unsuccessful machine deliveries, and legal conflicts. The breakthrough of the air-laid technology in Europe, North America, and the Far East took place in the 1980s.

CHAPTER 4

Figure 1. Papermaking processes.

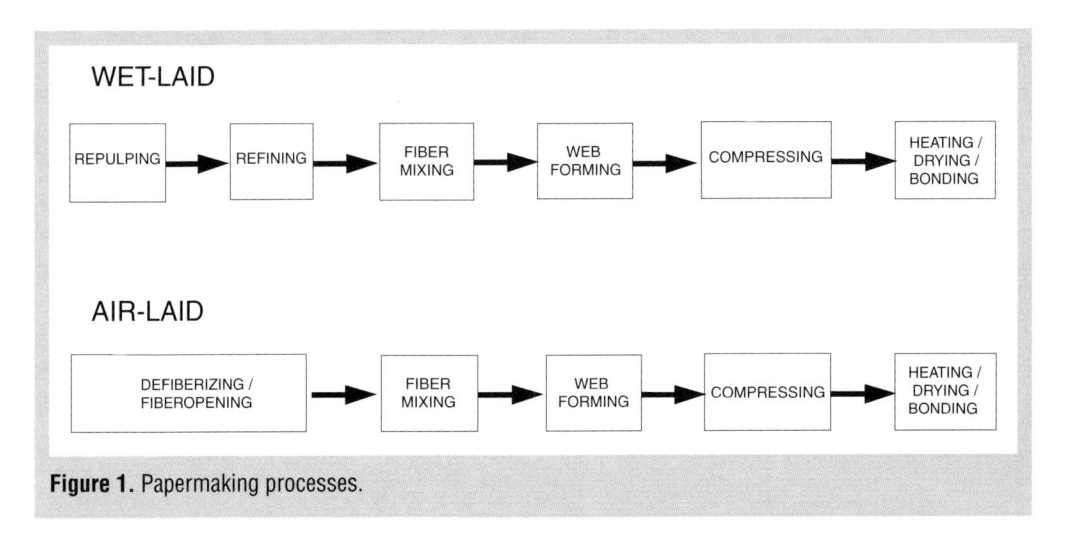

2 Process

Unlike the "normal" paper process, air-laid paper does not use water as the carrying media for the fiber (Fig. 1). Fibers are carried and formed to the structure of paper by air. The binding in normal paper takes place during drying by a binding mechanism created by the cellulosic components of the wood fibers. In air-laid paper (Fig. 2), the binder has to be applied in the form of spray, foam, or additional fibers or powders, which can be activated and cured by heat.

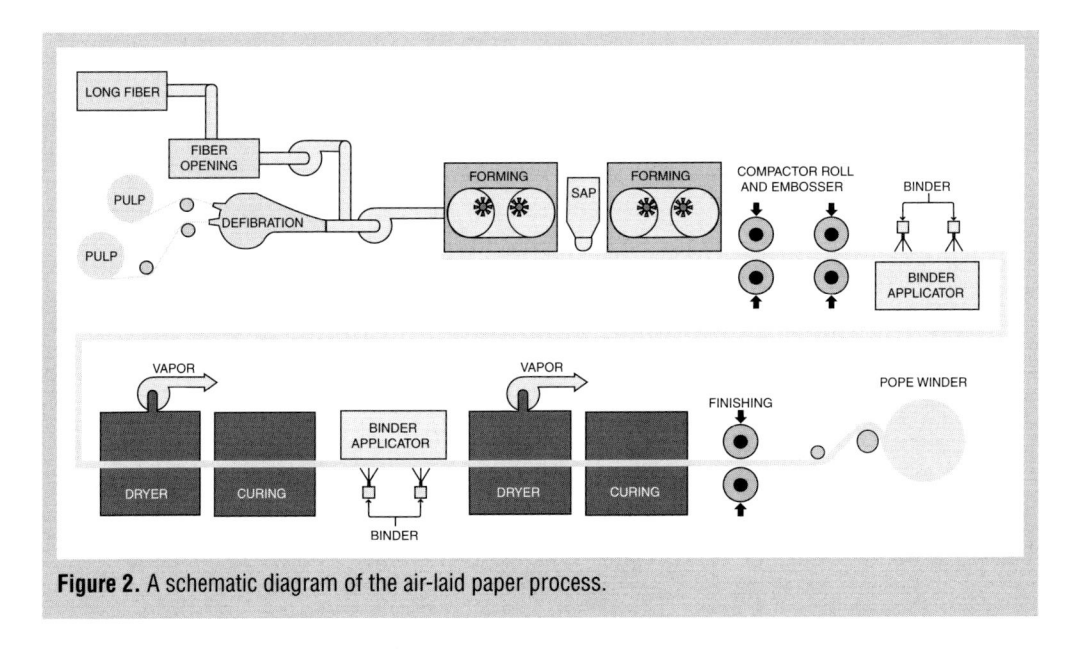

Figure 2. A schematic diagram of the air-laid paper process.

As described earlier, the air-laid paper is made of wood fibers. Long-fibered softwood pulp in roll form (fluff pulp) is the main raw material consisting about 85% of the weight of the basic paper. The other 15% in the so-called latex-bonded paper (LBAL) is typically EVA (ethylene vinyl acetate) latex. The other latex types are acrylic or SB (styrene butadiene) types normally used in different applications and industries including "normal" paper industry.

The alternative way of binding the fibers into the paper, or more widely into fiber web, is to mix into wood fiber matrix man-made fibers or powders that can be thermally activated. This type of paper is called thermobonded paper (TBAL).

Today most of the machines are hybrid machines, which can manufacture both types mentioned above or the combination of those so-called multibonded paper (MBAL).

During the years of the development of applications and technology, different kinds of new functional raw materials have been introduced into the paper either in fiber, granular, or liquid (dispersion) form. Among those, superabsorbent granules (SAP) in absorbent hygiene products are widely used.

3 Product characteristics and applications

Compared to normal wet-laid paper and tissue, air-laid paper is very bulky, porous (lofty), and soft. High porosity combined with hydrophilic wood fibers, gives air-laid paper good water and water solution absorption characteristics. Its strength, both wet and dry, can be modified by the amount and type of binder used. Compared to wet-laid tissue, the strength properties of a typical LB air-laid paper are clearly outperforming.

The above-mentioned characteristics at the beginning of the history of air-laid paper led to the idea of using it as an absorption media and for wiping purposes. Industrial, institutional, household, and hygiene wet wipes still comprise one of the major groups of applications. The recent development of the baby wipes sold either under the brand name of a baby diaper or a distributor's private label has increased the usage of air-laid paper remarkably.

During the 1990s, the trend in personal hygiene products, especially in feminine hygiene has been thinner products. This has resulted in the use of air-laid paper as the raw material in these products: sanitary napkins and panty shields. The addition of SAP powder into the paper gives extremely high absorption properties to the end products. The ease of converting such a prefabricated absorption material improves the productivity of the converting lines tremendously.

The similar trend is expected to occur in disposable diapers and adult incontinence products in a few years as well. That will mean a giant step in air-laid business.

The textile-like drape of air-laid paper has made it possible to use it as a raw material for disposable tabletop products, such as napkins and tablecloths. The paper has good printability and can be supplied fiber-colored in several colors. That gives luxury outlook to napkin designs and allows them to outperform multilayer tissue products.

CHAPTER 4

Main characteristics of air-laid paper are:
- Soft, does not scratch
- Non-linting, no dust, no static
- High absorption rate and capacity
- Strong even when wet, can be rinsed and reused
- Clean, hygienic, can be sterilized
- Does not irritate skin, produced of FDA-approved raw materials
- Textile-like surface and drape
- Can be dyed, printed, embossed, coated, made solvent resistant

Most common uses are presented in Table 1.

Table 1. Typical products.

Feminine hygiene	
Product 105 g/m^2 TB	
Thickness, 1-ply	1.3 mm
Tensile strength, MD	0.20 kN/m
Absorption capacity, free swell	26 g/g
Absorption capacity, under load	23 g/g
Table top	
Product: 60 g/m^2 LB	
Thickness, 1-ply	0.45 mm
Tensile strength, MD	0.34 kN/m
Wet strength, MD	0.18 kN/m
Wet wipes	
Product: 65 g/m^2 LB	
Thickness, 1-ply	1.3 mm
Tensile strength, MD	0.28 kN/m
Wet strength, MD	0.18 kN/m
Absorption capacity, free swell	13.0 g/g

Health care products	
Product: 60 g/m^2 LB	
Thickness, 1-ply	0.49 mm
Tensile strength, MD	0.37 kN/m
Wet strength, MD	0.22 kN/m
Industrial wipes	
Product: 53 g/m^2 LB	
Thickness, 1-ply	0.62 mm
Tensile strength, MD	0.48 kN/m
Wet strength, MD	0.26 kN/m

CHAPTER 5

Specialty papers

1	**What is a specialty paper?**	**101**
1.1	Economy	101
1.2	Life cycle	102
2	**Characteristics**	**102**
2.1	Mechanical strength	103
	2.1.1 Wet strength	103
	2.1.2 Elasticity	103
	2.1.3 Stiffness	103
	2.1.4 Abrasion resistance	103
2.2	Mechanical structure	103
	2.2.1 Uniformity, formation	103
	2.2.2 Pore size distribution	104
2.3	Absorption	104
2.4	Functional characteristics	104
2.5	Electrical characteristics	105
2.6	Cleanliness	105
2.7	Appearance	105
3	**Specialty grades**	**105**
3.1	Electrical papers	106
	3.1.1 Cable paper	106
	3.1.2 Capacitor tissue	107
	3.1.3 Transformer board	108
	3.1.4 Conductive paper	108
	3.1.5 Creped papers	108
3.2	Absorbent lamination papers, furniture papers	108
	3.2.1 Absorbent kraft	109
	3.2.2 Decor paper	109
	3.2.3 Overlay	110
	3.2.4 Pre-impregnated foils	110
	3.2.5 Laminating kraft	111
3.3	Filter papers	111
	3.3.1 Filtration mechanisms and paper structure	112
	3.3.2 Engine protection	113
	3.3.3 Laboratory and diagnostics	114
	3.3.4 Food and beverage	114

CHAPTER 5

	3.3.5 Coffee filters	114
	3.3.6 Tea bags	114
	3.3.7 Vacuum cleaner dust bag	115
3.4	Special strong papers	115
	3.4.1 Abrasive base	115
	3.4.2 Spinning kraft	116
	3.4.3 Hunting cartridge paper	116
3.5	Release papers	116
	3.5.1 Pressure-sensitive labels	117
	3.5.2 Posters, signs, and manual labels	118
	3.5.3 Vinyl casting	118
	3.5.4 Cover strips	118
	3.5.5 Masking tape	118
3.6	Copy and imaging papers	118
	3.6.1 Photographic paper	118
	3.6.2 Blueprint diazo paper	119
	3.6.3 Carbonless copy	119
	3.6.4 Thermal printing paper	120
3.7	Packaging	121
	3.7.1 Sack kraft	121
	3.7.2 Flexible packaging	122
	3.7.3 Label papers	124
3.8	Specialty printing	125
3.9	Special fine papers	126
	3.9.1 Security papers	126
	3.9.2 Artists' papers	126
	3.9.3 Tracing paper	127
	3.9.4 Plotting paper	127
	3.9.5 Luxury stationery	127
3.10	Building papers	128
	3.10.1 Wallpaper	128
	3.10.2 Barrier papers	128
3.11	Cigarette papers	128
3.12	Various functional papers	129
	Further reading	130

CHAPTER 5

Paul Olof Meinander

Specialty papers

1 What is a specialty paper?

All paper grades are specialized – the commodity grades, actually more specialized than most so-called specialty papers. In any event, because printing and writing are the most common uses for paper, as well as certain hygienic and packaging grades being produced in great volumes, it is adequate to define specialty papers by exclusion. Thus, a specialty paper would be one that does not primarily serve as information or print carrier, as protection, or as fluid absorber.

Specialty paper must also fill the criteria of having specific characteristics required for the specific uses; for instance, it does not necessarily take a specialty paper to make confetti. The distinction between specialty and commodity is thus very subtle and subjective.

1.1 Economy

A specialty paper is typically a paper with a specific feature. It is also typically only a minor component in the final product. In most cases, the share of the paper in the cost of the final product is small, whereas its importance for the function or quality of the final product is big. This means that the price elasticity of specialty papers is good. The user is prepared to pay a premium price for the product, and the threshold for changing suppliers is high.

Compare a magazine printing paper with a filter paper. The paper is a major cost item for the magazine publisher; it has a certain, but not decisive, influence on the value of the magazine. The paper used in a filter cartridge of an earth-moving machine is a minor cost for the filter maker and negligible for the contractor, but largely determines the function of the filter – and a failure might cause a disaster. Needless to say, the maga-

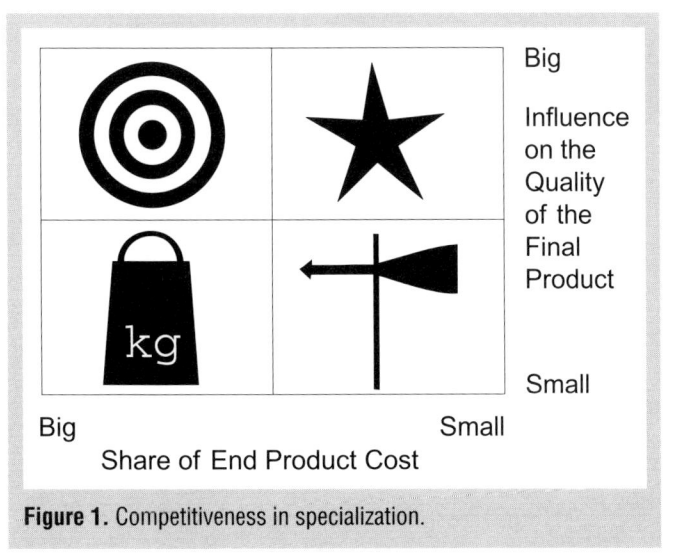

Figure 1. Competitiveness in specialization.

zine is more price sensitive, and the entry barrier for a new entrant to the filtering is higher. Figure 1 categorizes the competitive logic according to the cost and quality influences. Here the filter paper would be a star product, the magazine paper a weight. A wrapping paper may, but does not necessarily, behave as a weather vane, and a siliconizing base paper (which is a significant cost item) is purchased with great concern and precision.

Specialties are also, typically, minor quantity products and may be manufactured on small, less cost-effective paper machines, than typical commodities. The field of specialty papers is very fragmented; each use needs a specific grade.

Due to this fragmentation and complexity, the specialty papers can be dealt with only on a very general level and on the base of representative examples. A full and comprehensive description is not possible within the context of this book.

We, therefore, first look at some of the paper characteristics making a specialty paper, then view a few representative groups of specialties.

1.2 Life cycle

Many paper grades start their life cycle as specialty papers. As they grow, they change character to become commodities. This is the case, for instance, with laser printer and copy papers, which at the early stage of technology were veritable specialties. A decade later the bulk of these have to be regarded as commodities. As the volume grew, copiers and printers were developed for accepting a general paper rather than a customized one.

Some grades grow and then decline in use as other materials develop which can be substituted for those grades. Electrical insulation papers are an example of this type of a development. Synthetic materials are increasingly taking their place in power cables. Telephone cables have increasingly changed technology but are now dominated by optical cables.

A general trend in the industry is to search for ever more specialized products for older, outdated paper machines. These marketing dominated, small market volume products suit a smaller production scale. However, the entry barriers are high. An extensive product development is needed, often the culture of the whole organization has to be changed, and – especially for the "star" products – the acceptance process is very slow.

2 Characteristics

On a general level, a specialty paper typically has a set of characteristics, needed for a particular use. These may be very common ones, familiar for commodity papers, but more pronounced or with a closer tolerance. Such characteristics include strength, thickness, porosity, and absorption.

They may also be very particular characteristics, such as electrical conductivity, pore size distribution, resistance to certain chemicals, chemical reactivity, light proofness, heat resistance, and cleanliness.

Specialty papers

2.1 Mechanical strength

Even though every paper requires a certain mechanical strength, this demand can be extreme for certain specialties. Toughness to an extreme degree is required by spinning krafts, cartridge paper for hunting cartridges, abrasive base papers, and certain wrapping papers. The strength then mostly is required as tensile strength in one direction of the paper, whereas the other direction requires a maximum elongation to break.

The strength is obtained by using tough, long fiber pulp – sometimes combined with a certain portion (20%–25%) of shorter fiber for an improved formation and uniformity, avoiding weak spots. The stock is refined to a rather high degree, 40° Schopper-Riegler or more.

The MD/CD ratio is typically quite high, two or even more for maximizing the machine direction strength. This can at best be achieved by using vat formers that allow high dilution to be used. For thinner grades, fourdrinier formers are best suited.

2.1.1 Wet strength

Many papers having to be impregnated, or used in wet environment need a wet strength treatment. Wet strength is obtained by addition of wet strength resins to the pulp. These are often slowly curing, and the paper will reach full wet strength only a couple of days after production.

2.1.2 Elasticity

The required elasticity or elongation to break is mostly obtained by a high MD/CD strength ratio, which translates into a high cross machine elongation. This is further obtained by letting the paper shrink in machine direction during drying, causing micro creping in the fibers. Long and loose draws between the drying cylinders, loose drying fabrics, and a low freeness promote this.

2.1.3 Stiffness

A right degree of stiffness is critical for the function of papers that need to be fed through processing machines, like copiers, fax machines, or banknote counters.

2.1.4 Abrasion resistance

Value and security papers, which need to be handled many times during their life, require good resistance to abrasion or wear.

2.2 Mechanical structure

2.2.1 Uniformity, formation

Uniformity is important for most papers but, in cases where a paper is treated with huge amounts of water and dries free of tension, this becomes critical. An example of this is abrasive base papers, which are supposed to stay flat and uniform also after application of the grit and drying in free air.

Luxury stationery or fine printing papers similarly require an extremely good formation, specially when watermarked.

2.2.2 Pore size distribution

Porosity in general is not enough to characterize the holding or barrier characteristics of certain specialties. Filter papers and surgical sterilization papers require calibrated maximum pore sizes, which can be determined by covering the paper with a liquid and pressing air through the paper from the bottom side. Depending on the surface energy of the liquid, the pressure needed for passing the wet paper is determined and it gives a measure of the biggest pores in the paper. Measurement is made for the first bubble to pass and for the pressure causing a uniform "boiling."

2.3 Absorption

The absorption or repellence characteristics of the paper, which are also important when printing, are critical for certain specialties. Application of a waterous coating or binding agent can cause the fiber structure to swell and deform if water is excessively absorbed by the web. At the same time, it may be critical that the paper surface is wetted for getting sufficient anchorage of the coating.

For instance, when siliconizing a paper, the paper must be completely covered by a silicon film. While silicon is expensive, the film must be as thin as possible. Therefore, the paper must be well wetted by the silicone emulsion, but must not absorb it. A careful combination of refining, calendering, and coating produce the right mixture of characteristics.

For securing adhesion, the paper surface in some cases – like for critical polyethylene coating – might have to be corona treated before coating. Corona treating involves subjecting the paper to electrical discharge, modifying the surface structure.

Paper grades to be impregnated, on the other hand, require the right absorption of the impregnating liquid. For instance, blotting paper should absorb as much and as fast as possible. With products such as diapers, the rate and amount of absorption can be increased by adding special super-absorbents to the web.

2.4 Functional characteristics

Dimensional stability, that is the sensitivity of the paper to changes in humidity, is critical for papers that have to be laminated or collated together (e.g., copying papers in a set of forms). Cellulose fibers swell in their thickness direction, and are rather stable in their length direction. Therefore a "square" sheet, that is one which has the fibers randomly oriented, is more stable than one with more oriented fibers, i.e., having a higher MD/CD tensile strength ratio.

Over-drying the web can "kill" it, decreasing its sensitivity to humidity changes. The paper can be stabilized by over-drying before a subsequent treatment in a size press or a coater. For most papers where dimensional stability is critical, it is important to have the paper in the correct humidity and in equilibrium with the environment when using it.

The optical requirements of specialty papers can depend on their use. Certain flexible packaging or envelope window papers, as well as label backing release papers need to be transparent.

2.5 Electrical characteristics

Papers used in electrical cables or apparatuses require specific characteristics of conductivity or insulation properties. These are obtained by using very clean pulp, which is free of electrolytes. For the most demanding uses in high tension cables or capacitors, the paper is made by using distilled or deionized process water.

Other applications, on the contrary, require a certain conductivity that can be obtained by using carbon black as a filler or by treating the paper with an electroconductive resin in the size press.

2.6 Cleanliness

Cleanliness is important, but not only for the appearance of the final product. Chemical cleanliness is, obviously, important for diagnostic or chemical reagent carrying papers. Good cleanliness is required for optical character reading (OCR) grades. For photographic papers, any trace of radioactivity is detrimental and decor papers are most sensitive to dirt.

Cleanliness driven to an extreme is thus a characteristic of certain specialty papers, requiring special equipment as well as a specific environment in the production mill.

2.7 Appearance

Finally subjective properties like touch or appearance may characterize a specialty paper. For instance, touch is a factor for security and high class stationery papers.

Colored papers generally require a good uniformity of shade. For decorative laminating papers, where impregnation deepens the color and the end product is highly visible, the required shade accuracy is extreme.

The structure or texture of the sheet can be a purely aesthetic issue for papers like luxury stationery. It can become more important with paper requiring half-tone watermarking, security markings, optically active mingling, and such as is used in security papers.

Appearance features may require specialization as well as more technical characteristics.

3 Specialty grades

In the following, a number of typical specialty paper grades are reviewed. As mentioned, the field of special applications for paper is large and fragmented, so only a representative sample can be dealt with. The sample should, however, give a good idea about the versatility of paper and the variety of products and production methods.

The products are looked at according to their end use and the requirements set by the end use. The next step is to see how the requirements are met by means of stock composition, basis weight, and manufacturing techniques. Finally the development trends and outlooks for the product are considered.

CHAPTER 5

3.1 Electrical papers

Pure cellulose has outstanding electrical properties. It is a good insulation material and is also polar, having a proper dielectric constant. Paper has therefore traditionally been used for electrical insulation in many applications. The electrical insulation papers are typical technical specialties and therefore have experienced a golden age. However, technological development has now surpassed them, and they are gradually being replaced by other materials.

The more demanding the application, the cleaner the paper must be. It is therefore common to run the paper machine with chemically deionized or even distilled process water when producing the higher grades of electrical paper. Various electrical papers are summarized in Table 1.

Table 1. Electrical papers.

Electrical papers	Thickness range (µm)	Demand on process water	Specific characteristics
Telephone cable	50–70	Normal	Cleanliness
Power cable	60–190	Normal	
High tension cable	55–190	Ion free	
Capacitor tissue	5–12	Ion free	Density, free of holes
Conductive	100–120	Normal	Carbon black
Transformer board	Up to 2000	Varies according to use	Creped

3.1.1 Cable paper

Electrical cables are categorized by the tension and the current used. The lowest tensions and currents are normally found in telephone cables, transmitting electrical signals. The highest tensions and high currents are found in high tension power cables.

Figure 2 represents the construction of a paper isolated cable.

Cable papers are produced from unbleached kraft pulp, whereby chemical and mechanical cleanliness are important.

Telephone cable paper

In telephone cables, a huge number of threads are individually isolated by paper. In order for the cable not to become excessively thick, the paper has to be thin (30–40 g/m² is normal).

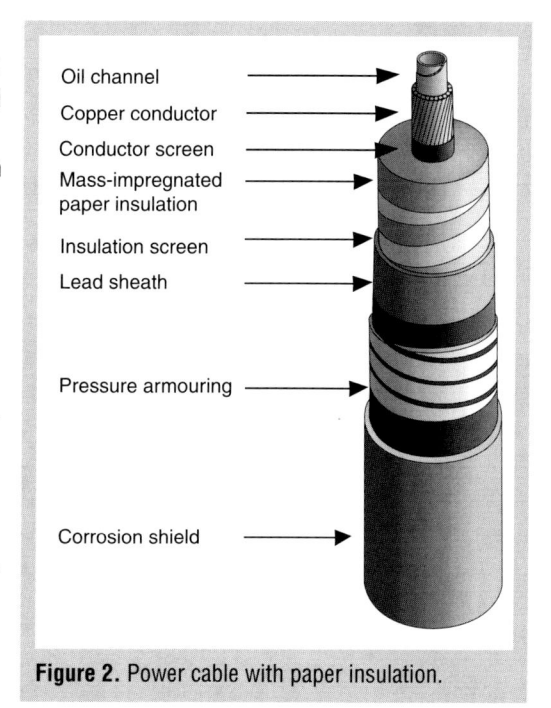

Figure 2. Power cable with paper insulation.

Specialty papers

The strands of the cable are wound in high-speed spinning machines, requiring a strong, elastic, uniform paper free of holes or debris.

Before spinning, the paper is sliced into narrow (about 5–20 mm wide) disc-shaped rolls, with a length of up to 4–6 km, which requires a good thickness profile. Electrical characteristics are less critical.

This machine finished paper has generally been made on small fourdrinier machines of good quality, well-delignified softwood kraft. The market is now declining because plastics are used for insulation and optical cables are replacing the electrical ones in long distance applications.

High tension power cable paper

Sea cables for power transmission at very high tensions up to 400 kV or more are a very demanding application for paper. The paper has to be very clean for avoiding leaking currents, and the paper is produced with deionized water.

This paper is in the range of 65–155 g/m^2 and is mostly produced on twin-ply paper machines. Due to the high degree of specialization and the need for ion-free water, the scale of production can only be small.

This type of cable paper is a typical product where the technological entry barrier is very high. The paper makes only a minor share cost, but is important for the functional characteristics of the cable. The two wires improve formation and eliminate the risk for holes.

A particular advantage of using paper in sea cables is that, in case of a leakage, the paper will swell and prevent water from flowing along the cable.

Power cable papers

Between the extremes of high tension power and telephone cables, there is a wide range of intermediate types of cable papers. However, plastic materials are also being substituted more and more for these papers.

3.1.2 Capacitor tissue

Capacitor tissue used to be a real specialty, but has now largely been replaced by metallized plastic films.

This paper is mainly used in power condensers for compensating the inductive component in industrial power consumption. Before the era of plastics, capacitor tissue was used in most capacitors.

Capacitor tissue is made of super clean unbleached kraft pulp, using deionized water. The capacitance of a condensator is inversely proportional to the thickness of the insulating medium. Therefore capacitor tissue is made as thin as possible.

The basis weight typically ranges from 6 to 12 g/m^2 and the paper is supercalendered to a density of 1.2 g/cm^3. The grade represents an extreme of thinness and can be made only on small, relatively slow paper machines, with a trim of 4 m or less. The stock is refined extremely far, to more than 90° Schopper-Riegler.

An absence of pinholes is vitally important, and the stock must drain very slowly. Also production speeds are low.

CHAPTER 5

3.1.3 Transformer board

For uses where an insulation structure is needed, thick transformer boards are used. Transformer board is made in a very traditional way, where the board is built up on forming cylinders, with simultaneous pressing. When the board reaches a desired thickness, it is cut off the cylinder, and brought to drying as sheets, with the length of the sheet corresponding to the circumference of the drum.

The wet sheets are fed to drying in separate drying machines, impregnation, calenders, finishing presses, and so on. For ultra high tensions, they can be laminated together to form thicker boards.

Transformer board is a typical, small volume specialty, which has mostly been replaced by other materials, but is still surviving for certain particular uses.

3.1.4 Conductive paper

Static electricity must be discharged in order to avoid gleaming, in which case an electroconductive paper may be used. This paper is traditionally made by loading with carbon black, which is a conductive pigment.

Carbon black is very aggressive and penetrant, so cleaning up after the process of using carbon black in the system is laborious. It can take up to a couple of weeks of cleaning before the system is clean enough for making white papers.

3.1.5 Creped papers

As well as for insulation, conductive papers are also applied manually. This can be facilitated by creping the papers, which can be made on off-line creping machines, which permit a degree of creping up to 100% elongation.

3.2 Absorbent lamination papers, furniture papers

Laminates for furniture, decoration panels, floors, and so on are becoming a growing use of very special papers. These are normally laminated onto a particle- or fiberboard, giving a good-looking resistant surface.

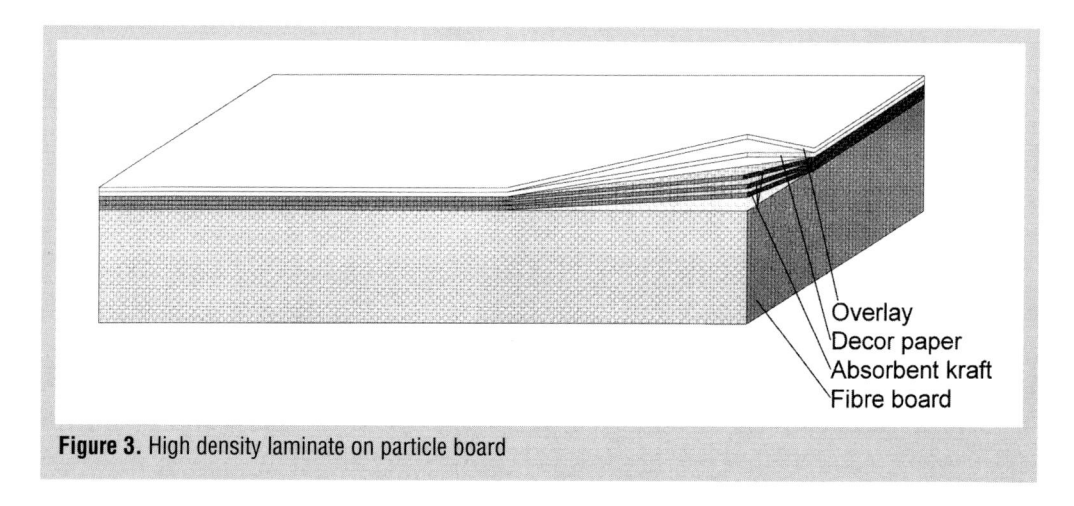

Figure 3. High density laminate on particle board

Specialty papers

A laminate consists of multiple layers, each having its own distinct function (Fig. 3). It is impregnated by a resin, mostly melamine, which is cured to form an inert, hard composite with the fiber structure of the paper. The highly visible use of the laminates, as working tables, cupboard fronts, or floors sets very high demands on the appearance of the laminate, which must be clean, have the correct light-resistant shade and surface structure, and be resistant to wear and scratching.

The laminate is normally built on a body of absorbent kraft, which may be followed by a light underlay paper and then a decor paper giving the appearance of the laminate, and finally an overlay paper giving wear resistance.

The laminate is built so that the plies are impregnated with resin, after which they are cut into sheets and pressed together in a high-density press, with separating plates between the laminate boards. The pressure is high enough for pressing out any gases contained in the laminate, which thus becomes a solid, poreless structure.

3.2.1 Absorbent kraft

The absorbent kraft paper is basically a normal kraft, with controlled absorbency, i.e., a high degree of porosity. It is made of clean low kappa hardwood kraft and has to have a good uniformity. It should have a reasonably good formation and good profile for the uniformity of the laminate.

Absorbent kraft is an 80–120 g/m^2 paper and, for a laminate, 2–4 plies are used.

3.2.2 Decor paper

The decor paper is the most critical of the laminate papers. It is the visible paper giving the appearance to the laminate.

The impregnating resin and cellulose have roughly the same refraction index so that the fibers in the decor paper appear transparent after laminating. This means that, at laminating, the shade of the paper deepens dramatically and actually only the pigments and dyestuffs remain visible.

The human eye is very sensitive to color differences, and the decorative laminates used in large, exposed surfaces make the dyeing of decor paper extremely difficult and critical. In order to cover the body of the laminate, a huge amount of pigments is required, and only pigments like titanium dioxide, which have a very high refraction index, are efficient. So decor paper normally contains at least 20% and even up to 40% or more of titanium dioxide or other special pigments.

Decor paper is made for white or colored decor, often also imitating wood finishes. There are therefore, typically, many variations of colors, and even the white is produced in a number of different shades. Making the colors is made more difficult due to the fact that they can be seen only after laminating. So, when adjusting the color, a sample must be drawn from the paper machine and laminated in the laboratory, before the result can be judged.

When the lots are small and color changes frequent, the capability of changing shades becomes a major cost factor in producing decor paper. Therefore the paper is preferentially produced on small paper machines, which must match the widths of laminators on the market.

CHAPTER 5

The production further requires an extreme cleanliness of the system and the dryer part of the machine. Because impregnation makes the sheet structure transparent, any dirt in the paper will be highly visible.

All in all, this is a paper grade that requires particular skills and specific equipment and, in spite of its apparent suitability for many small machines on the market, the entry barriers are very high.

3.2.3 Overlay

The overlay paper is a 20–50 g/m² open sheet made of pure cellulose. It thus becomes transparent after impregnation, letting the appearance of the decor paper come through.

For flooring, which requires particularly high resistance to abrasion, a hard pigment like corundum may be added to the overlay. In this way, the surface can also be made less slippery.

Special requirement for overlay is high uniform porosity, a reasonably good formation, cleanliness, and light resistance. Thus it must be made out of very well delignified pulp.

Table 2 summarizes the typical papers of a high-density laminate.

Table 2. The papers of a high-density laminate.

High-density laminate	Basis weight/thickness	Critital characteristics
Overlay	20–50 g/m²	Cleanliness, transparency as impregnated
Decorative	70–150 g/m²	Opacity, cleanliness, shade as impregnated
Underlay	100–150 g/m²	Lightness and opacity as impregnated
Absorbent kraft	80–120 g/m²	Porosity, absorbency
Particle board	10–30 mm	Mechanical stability (no paper)

3.2.4 Pre-impregnated foils

Pre-impregnated foils are used where the surface requirements for resistance and closedness do not require more expensive high-density laminates. They are typically used for cabinet interiors and sides and for first home, kids', and youth furniture.

Pre-impregnated foils are not fully impregnated; therefore, the quality demand on pre-impregnated foils is not as extreme as for decor papers. The final shade may also be seen shortly after production, so color adjusting is not quite as time consuming as with decor papers.

The shades are mostly various shades of white, or brown for printing wood textures on the paper. Wood shades have to be matched to different species. Printing to imitate the wood finish is made by offset or screen printing whereby a three-dimensional effect is possible.

The foils are made in basis weight range, 80–120 g/m². A thicker variation, 200–400 g/m², is used for covering edges of the panes. The furnish used is a mixture of hardwood and softwood for optimizing runnability and formation. Inorganic pigments, mostly iron oxides, are used for coloring.

Specialty papers

Impregnation can be made on the paper machine, using a modified size press. Resins for impregnation are of different types, depending on the degree of stiffness required for the application of the foil. A very stiff foil gives a better coverage of the structure of the underlying particle board pane, but is also very brittle and delicate to handle.

A particular problem when producing pre-impregnated foils has been the noxious formaldehyde gases that develop during curing of the resins used. This problem has been at least partly avoided by the use of low formaldehyde melamines or ureic resins.

3.2.5 Laminating kraft

A still cheaper material for lining particle boards is a laminating kraft paper. A natural, smooth, and dense kraft paper can be laminated onto the particleboard and, consequently, coated or varnished to give the surface washability and resistance dependent on the coating used.

A kraft paper may also be applied on the rear side of laminated boards, for compensating the tensions created by top-side lamination.

3.3 Filter papers

There is a vast range of filter papers for various purposes. The use of porous paper for filtering and separation ranges from the use of filter cartridges for engine protection to dust pouches for vacuum cleaners and air conditioning, and from tea bags and coffee filters to reagent carriers and filter discs for laboratory use.

Their common denominator is the requirement for a controlled, high porosity. Depending on the intended use, additional required characteristics are the resistance to different media or temperatures, stiffness, and cleanliness.

Specific characteristics of filter papers are their resistance to flow, their filtration efficiency, and their dust-holding capacity. These characteristics vary over a wide range according to the application. They are all functions of the porosity and pore size distribution of the paper.

Filter papers are also manufactured from a wide range of fiber materials. Coarse long fiber pulps for filter paper use may be specially treated for retaining a high porosity. On the other hand, fine short fibers like eucalyptus or special fine long nonwood fibers like abaca or Manila hemp are used. Due to their partially irreversibly dried, curled structure, flash dried fibers are often used for making filter papers. Also inorganic fibers or synthetic fibers are used in filtration media, but these are too specialized for being considered here.

Because the porosity is mostly higher than for standard papers, the resistance to flow has to be measured as the pressure drop over the paper at a certain flow of medium through the paper, or inversely as the flow at a certain pressure drop.

Filtration efficiency and dust-holding capacity are determined in test rigs, where the medium is contaminated in a controlled way by a powder of known characteristics. A particle counter determines the number of dust particles in a certain share of the downstream flow through the paper sample.

CHAPTER 5

The dominating pore size can be determined by a bubble test, fixing the sample to the bottom of a testing bowl, covering it with a wetting fluid of known surface tension, and pressing air through the paper from the bottom side and registering the air pressure. The air pressure at the first bubbles passing the paper gives an indication of the biggest pore size; the pressure when the fluid begins to "boil" indicates the dominating pore size.

Pore size distribution can be investigated by pressing mercury into the paper. The amount of mercury entered at a given pressure gives a picture of the volume of pores of a certain diameter.

3.3.1 Filtration mechanisms and paper structure

The filtration itself can take place according to two main mechanisms (Fig. 4). In-depth or volume filtration, the dust to be withheld enters the structure of the paper and is trapped between fibers in the porous structure. The filter is gradually filled up with dust, and its permeability decreases until the filter has to be changed.

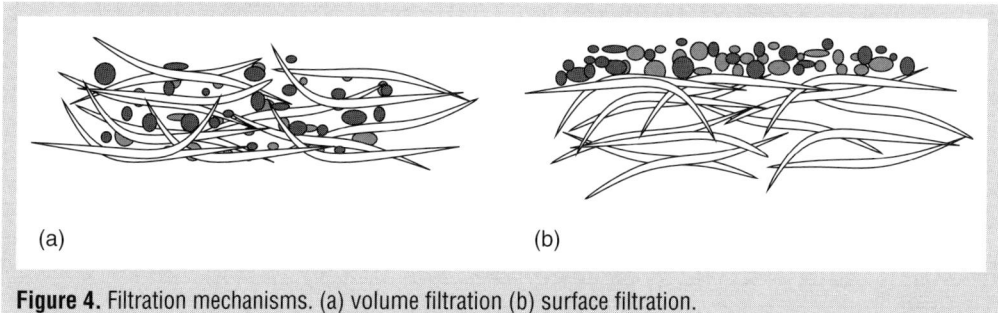

Figure 4. Filtration mechanisms. (a) volume filtration (b) surface filtration.

In surface or cake filtration, the dust is stopped at the surface of the filter where it builds up a layer of dust (the cake), which in itself acts as a filter. The resistance to flow increases with the increasing thickness of the cake. At the same time, the filtration efficiency also increases. In certain cases, removal of the cake can extend the lifetime of the filter.

For volume filtration, the filter should be built so that the entry side of the filter is more porous than the exit side. Thus the entry side permits a part of the dust to pass toward the denser parts of the filter, and the dust builds up through the whole volume of the filter. If the entry side was the tighter, it would clog faster and the capacity of the body and exit side of the paper would be unused.

Even in the case of surface filtration, the filter slowly fills up, which limits the lifetime of the filter. It is therefore better that the entry side forms an efficient barrier, letting as little dust as possible enter into the structure of the filter. The body of the paper then functions as a mechanical structure carrying a thin filtering barrier. The pore size of this barrier then should be so small that the dust is efficiently prevented from passing.

The permeability and pore-size distribution is not uniform through a paper. It can also be influenced by technologies used in the production. A typical paper sheet made on a fourdrinier has a certain two-sidedness due to the washing out of fines at the wire side of the paper. On the other hand, the mechanisms of draining cause the fibers at the wire side to form a more closed structure than on the top side of the sheet. If the stock contains only little fines, as is the case for many filter papers, the latter effect dominates, and the wire side of the sheet is denser than the top.

3.3.2 Engine protection

Automotive filter papers make a big and highly specialized family of papers. They can be split into three main categories: papers for combustion air, lubrication oil, and fuel.

The filter papers are all transformed into filter cartridges, which are then fitted to the engine. The construction of the cartridges mostly requires that the paper is stiff enough for being self supporting and that the paper is embossed in a way for improving stiffness and forming channels for the flow of the medium to be filtered. The impregnation also stabilizes the paper against moisture and glues the fiber structure adding strength. Figure 5 shows a typical filter cartridge.

Air filter

Filter paper for air is the most open of the engine protection filter papers. Due to the low viscosity of air, the forces acting on the dust particles are lower than the the papers for oil or fuel.

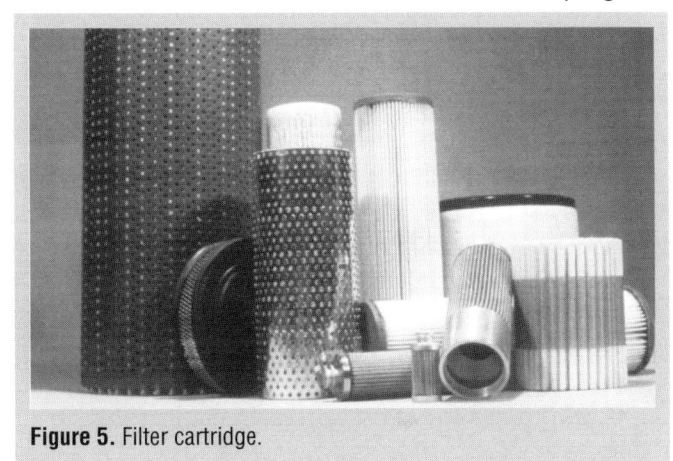

Figure 5. Filter cartridge.

The paper is 100–200 g/m² and very porous. This may be obtained by using particularly long fibers, which are partially mercerized, i.e., treated with acid in order to retain a cylindrical, uncollapsed form.

Often air filters for heavy equipment, like earth-moving machines, are categorized separately as *Heavy Duty* air filters. These have to function in environments with extremely heavy dust loads, and therefore also have a very high dust-holding capacity.

Air filtration can be volume or surface filtration. Heavy-duty filters work on the principle of surface filtration, and the cake can be rinsed off for extension of the life of the filter.

Oil filter

In lubrication oil, impurities are carried by a fluid with a viscosity much higher than that of air. For avoiding wear of the engine, the impurities permitted in the oil must be smaller than the thickness of the lubrication film in the engine cylinder – and as small as possible for the bearings. A cake would cause a high resistance to flow; therefore, lubrication oil is normally filtered in volume.

The oil filter papers are lighter and denser than the air filter papers. They are impregnated to resist high temperatures.

CHAPTER 5

Fuel filter

Diesel engines have very fine injection nozzles. Even small dust particles would block the injection. The fuel therefore has to be filtered for security. The fuel as such has already been filtered; the impurities contained are contamination from handling and transport.

The flow of fuel is also smaller than the flow of air or lubrication oil, so the filtration cartridges can be made more simply than those for air and oil.

The paper used for fuel filters is a creped paper with controlled porosity, which is pleated and wound to cartridges. The creping permits the fuel to be filtered to flow between the layers of paper.

The paper is made of a mixture of hardwood and softwood pulps, and creping is made wet on the last press of the paper machine or on a drying cylinder in the first part of drying. The basis weight of the paper is 50–80 g/m^2.

3.3.3 Laboratory and diagnostics

For laboratory use, filter papers are made in a variety of ways. Specific uses require specific papers, which can be made out of a wide variety of fibers, from acid washed wooden fibers to carbon or quartz fibers. The scale of production may be so small as to fit a paper machine of 50 cm width.

Finishing and confectioning is the bulk of production work, and the paper is distributed in small lots.

These papers may be impregnated by reagents for use in various detection tests, from pH to pregnancy or diabetes.

The complexity of distribution and confectioning makes these products world brands even at a small production scale. The entry barriers to the market are therefore very high. New analytic and diagnostic methods partially replace the laboratory filter papers in many areas.

3.3.4 Food and beverage

For food and beverages, filter papers and boards are widely used. Industrially filtration is mostly made in presses, where the product is filtered through boards or fibrous wads. The thickness of these boards provides very efficient filtration while also retaining germs. These boards that are up to 1 000 g/m^2, are die cut to plates fitting the press.

3.3.5 Coffee filters

Coffee filters are well known to most people. The paper is creped in order to let the coffee flow between the paper and the filtration funnel. The paper, about 100 g/m^2, is made of coarse long fiber, often from fast grown trees. Bleached and unbleached pulp is used.

3.3.6 Tea bags

Tea bags are made of abaca fibers, a very thin and long fiber hemp. The bag paper is very porous and thin, 12–20 g/m^2, and maintains extreme wet strength by heat sealing with a portion (<30%) of synthetic fiber.

3.3.7 Vacuum cleaner dust bag

Vacuum cleaner dust bags are another well-known household specialty paper. These papers need to retain the dust efficiently while having a porosity sufficient for passing the air from the vacuum cleaner. The paper, 100–150 g/m^2, is mainly made of softwood pulp; the porosity can be adjusted by adding hardwood.

3.4 Special strong papers

3.4.1 Abrasive base

Papers for abrasive base are an example of papers where the toughness is of great importance. They additionally require other characteristics.

The abrasive base paper is coated by an abrasive grit in a binder. It is used for belts in heavy grinding machines, as sheets for grinding by hand, in vibrating hand-held grinding machines, or as discs in rotary machines.

The paper must be strong enough to resist the forces in use, give a good anchoring of the grit, and suit the coating operation.

Coating and spreading the grit is critical, in that the abrasive paper is dried and the binding resin cured while hanging free in drying and curing ovens. During this operation, the base web gets wet and internal tensions are released. Therefore, the absorbency and uniformity of the paper have to be well controlled.

Abrasive base paper is made in different weights, from the lightest ones (70–100 g/m^2 grade A) to the heaviest (220–450 g/m^2, grade G). There are specific grades for wet grinding and for dry finishing by hand.

Abrasive base paper is a mature product with low growth. The high demands, the comparably low share of cost in the finished product make abrasive base a typical star product in the sensitivity matrix (Fig. 1).

Base paper for grinding belts

The high weight papers are made multilayer, either on vat formers or multi-fourdriniers. The trim width should, ideally, be equal to the width of the spreading machine in order to make symmetrical full width rolls. Drying has to be made cautiously to avoid dried-in strains in the paper. For releasing strains in the paper structure, the paper may be re-wet or sized in a size press.

The typical paper for heavy grinding belts is 200–250 g/m^2. The furnish is unbleached or semi-bleached tough long fiber pulp, with an addition of short fiber for improving the formation. Refining is modest, about 25–30°SR, as a compromise between strength and dimensional stability. Ply bond is improved by spraying starch or another binder between the layers.

Thinner papers are used for less demanding applications. They can be made on double-wire machines, or even with a single fourdrinier.

CHAPTER 5

Base paper for hand-finishing papers

Abrasive papers for finishing by hand are mostly used on complex surfaces and small pieces, which cannot be ground using a machine. The papers consequently need to adapt to these surfaces, which requires great flexibility.

One way to get this flexibility is to use latex as a binder in the very paper structure. Another is to wet the paper before use, which again requires a high degree of wet strength. One advantage with wet grinding is that the grit is rinsed by the fluid present when grinding. This permits an extended use of the paper as compared with dry grinding, which leads to faster clogging of the grit.

The hand-finishing papers are typically made on twin fourdrinier paper machines.

Sandpaper

Sandpaper stands for the simplest and cheapest of grinding papers. It has little to do with the real specialty papers and may be produced using a normal, good kraft paper.

3.4.2 Spinning kraft

An especially strong (about 40 g/m^2) kraft paper is still used for spinning paper strings for various purposes. This paper requires the best possible machine direction strength and cross machine elongation. This is caused by high fiber orientation. Its use is decreasing.

3.4.3 Hunting cartridge paper

Shot gun cartridges used to be, and to some extent still are, made of paper. This paper needs a high tensile strength in the machine direction, which is the axial direction of the cartridges. In the cross direction, the cartridge is supported by the gun-pipe, but a sufficient elongation is needed.

The body of the cartridge is wound of a kraft paper of 80–120 g/m^2, which is further covered by an intensely colored 60–80 g/m^2 sheet.

3.5 Release papers

Release papers are used as backing paper for self-adhesive labels. They are also finding use in packaging sticky materials and in industrial processes as casting papers.

The most common release material is silicon, which is applied in a thin layer on the base paper. Because silicones are expensive, the coating layer is applied as thin as possible (less than a micron thick). For good releasing, the silicon must form a uniform film without holes.

The capability of the base paper to support forming of this film, with a minimum of silicon, is the most critical specific property of a good siliconizing base paper. The paper needs to have a smooth and closed surface for the forming of a thin film, but also needs a structure compatible with the silicone for anchoring it to the paper.

Specialty papers

The silicon is applied as a solution or an emulsion, which is cured on the paper. The curing of the film is sensitive to inhibitors that might be present in the paper. Specific silicones therefore need specific formulations of the paper surface.

The surface is made by coating a paper with an open fiber structure or by supercalendering a glassine-type paper so that the fiber structure itself gets closed and forms a dense surface. Glassine is a paper, in which the pores are closed so that it forms a compact, transparent film of cellulose. For improved forming of the silicon film, the glassine papers are sized in a size press or a film press.

Siliconizing base used to be a specialty product produced on small 2–3.5 m wide machines. The growth of the market has outdated these machines, and the machine size currently used continues to increase.

3.5.1 Pressure-sensitive labels

Pressure-sensitive or self-adhesive labels are produced by die cutting the label paper in a laminate composed from top to bottom of label paper, pressure-sensitive adhesive, release coating, and base paper.

The die cutting requires that the caliber of the paper is exact enough for permitting the cutting tool to penetrate the label fully, but not to cut into the base paper. After die cutting, the grid between the labels is torn off, which requires neat cutting. The thickness profile of the base paper is therefore quite important (Table 3).

Labeling machines need to detect the exact position of the label for placing it correctly on the goods to be labeled (Fig. 6). This is often done by photocells through the paper. A sufficient degree of transparency is required from label backing paper, which is mostly a glassine type with a basis weight around 60 g/m^2.

Figure 6. Pressure-sensitive labeling.

3.5.2 Posters, signs, and manual labels

For labels that are not applied mechanically, signs, and posters, transparency is not required and a coated backing paper is mostly used. This 80–120 g/m² paper is made of bleached kraft pulp and is coated with a pigment coating of the right characteristics. The paper is in some cases produced as machine glazed and the coated surface is machine finished.

3.5.3 Vinyl casting

Synthetic leather is made by embossing a matrix of the leather texture into a siliconized paper and then casting a vinyl film upon this paper. The vinyl thus gets a leather-like texture. The same method can be used for other surface structures as well.

The paper is a 120 g/m² coated sheet, and the base paper is made with a high amount of short fibers for enabling precise embossing.

3.5.4 Cover strips

Various products – for instance, diapers and envelopes – have adhesive strips, which need to be covered before use. Here a thin siliconized cover strip of about 25–40 g/m² is used. For siliconizing, the base paper is coated by an extruded 10 g/m² polyethylene layer. The paper itself can be a machine glazed kraft paper.

3.5.5 Masking tape

Thin, 25–30 g/m² creped paper is siliconized on one side and adhesive coated on the other for making masking tape, which is used for covering unpainted parts when painting. The paper is wet creped on the last press or an early drying cylinder, and the crepe is stabilized in the dryer part of the paper machine.

Masking tape requires a creping grade of about 20%–25% and controlled absorbency characteristics. It is made with wet-strength sizing for stability in the coating process.

Table 3. Release papers.

Release base papers	Basis weight	
Glassine	50–150 g/m²	SC, high density, size press treated
Coated kraft	80–120 g/m²	Coated, kraft
Vinyl casting	120–150 g/m²	Coated, short fiber
Cover strips	25–40 g/m²	MG PE-coated
Masking tape	40–70 g/m²	Creped

3.6 Copy and imaging papers

3.6.1 Photographic paper

Paper for photographs must carry a uniform emulsion coating, it must resist the development solution in the development bath, it must be perfectly clean for a clean image, and it cannot contain any inhibitors to the photochemical process like iron, copper, or sulfur. It must even be free of radioactive traces, which cause photographic reactions and cause spots in the image.

Specialty papers

The paper, which must have a stable white color, is made of clean bleached pulp. (Even cotton or cotton linters are used for certain grades.) The pulp is refined for the best possible formation. In order to obtain a resistance against the reagents and rinsing water, the web is dip sized with gelatin, polyvinyl alcohol (PVA), polyacryl amide (PAA), and modified starch before calendering. Most photographic papers for color prints are extrusion coated with an opaque plastic film for improving impermeability.

Photographic papers are made in a lot of different variations for specific uses and make up a highly specialized business with high entry barriers.

3.6.2 Blueprint diazo paper

With the increasing use of computer-aided design and easy printing, traditional blueprint copies are losing ground.

Handling the large drawing copies, often in difficult environments, requires a good strength of the paper. Further coating and developing the image require a well-sized and chemically neutral, uniform surface. Also good opacity and brightness are required.

The paper is mostly 60–75 g/m^2 and is made of bleached kraft. (In earlier times, bleached sulfite was widely used.) It is rosin sized with the addition of wet-strength resin and is additionally treated in the size press.

3.6.3 Carbonless copy

Carbonless copy paper has largely substituted the use of carbon paper in multi-ply forms. With the dominance of word processors and nonimpact printers over typewriting, carbon copies have more or less disappeared from use in correspondence, memos, etc.

The dominating principle for carbonless copy is that an emulsion oil-carried color former is encapsulated in microcapsules applied as a coating on the backside of the copying paper. Through the pressure of writing, the microcapsules are broken, and the color former solution flows to wet the front side coating of a receiving sheet. The front side coating reacts with the color former, forming an image.

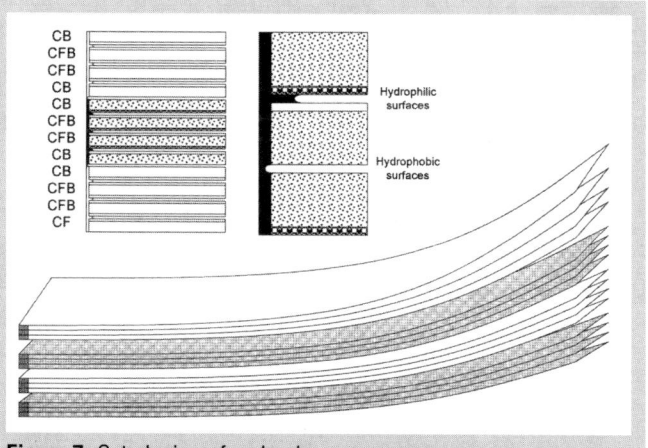

Figure 7. Set glueing of carbonless copy paper.

This principle was originally covered by a patent owned by National Cash Register (NCR) which then gave the name NCR (no carbon required) to the method. During the validity of the NCR patent, numerous other systems were developed and marketed. After the patents expired, the NCR principle was widely copied, and carbonless copy paper developed into a commodity with many suppliers. Now the market has matured.

CHAPTER 5

The sheets in a set of carbonless copy papers are named according to their coating and position in the set. The top sheet receiving the original writing is CB or coated back. The bottom sheet, which only receives an image, is CF or coated front. The middle sheets, which as well receive and pass an image, are CFB or coated front and back.

Carbonless forms are preprinted like normal forms – mostly in normal offset or web offset printing. By covering some fields by printing an inhibitor on them, copying can be selectively prevented.

Set gluing

For forming sets of forms, the printed forms can be collated in the required sequence and cut. When a watery glue is spread on the side of the pile of forms, it will enter between the CB and CF coated surfaces, gluing them together, leaving the uncoated top of the top sheet and bottom of the bottom sheets dry (Fig. 7).

This effect is obtained by making the coatings hydrophilic and hydrophobic, sizing the uncoated surfaces.

Continuous forms

The biggest volume of carbonless copy paper is used in continuous stationery form printers, where security and backup copies are required. The continuous forms are preprinted in web offset or flexo printing, after which the rolls are collated in roll collating machines. For this operation, an equal length of the webs is important. Dimensional stability, internal tensions, and – above all – a controlled humidity are required for this.

Production

Carbonless copy papers are mostly produced on normal fine paper machines, with the CF coating applied on-line with blade coaters. In order to preserve the microcapsules, CB coating is applied by air brush or other softer methods. Basis weights for CB are mostly around 50–70 g/m^2, which is lighter than for normal stationery papers to enable better pressure transfer. Base sheets for CFB are about 40 g/m^2 which, with coatings, results in about a 50 g/m^2 sheet. The CF bottom sheet can be made independently of pressure transfer, and often a heavier 70–120 g/m^2 sheet is used for giving stability to the form set.

Carbonless copy paper is partly made in colors for distinguishing the form layers. Coloring is then made by coloring the CF coating. This use is declining, though, because printing in different colors can produce the same distinction.

3.6.4 Thermal printing paper

Thermal printing is a simple, maintenance-free, and reliable printing method for minor printing works. Therefore thermal imaging has a widespread use in facsimile (telefax) machines, tickets, sale price tags, and label printing.

In a thermal printing paper, a thermosensitive wax separates the color former and the receiving reactive substance. When the surface is heated, the wax melts down, which allows the two color components to flow together, forming the image.

Specialty papers

Thermal printing paper is made in a wide range of qualities, the most common one being facsimile paper rolls. The base paper is mostly a 50–60 g/m² rather dense sheet of paper, which is surface sized for stability.

More sophisticated grades are made for labels and tickets where the thermosensitive layer is cover-coated for protection from the influence of moisture and mechanical scratching.

Thermosensitive paper is an example of a specialty paper with low entry barriers, where a fast initial growth and good revenues attracted a great number of competitors. This caused a heavy oversupply, price collapse, and shake-out competition until the market stabilized.

3.7 Packaging

Packaging can be classified as transportation, distribution, or consumer packaging. Transportation packages are mostly made of corrugated board and serve for protecting the goods during transportation. Distribution and consumer packages protect and preserve the goods, but also are a part of the product. They need to be attractive for promoting sales, carrying information about the product, and being practical to use.

Consumer packaging can be stiff boxes made of folding boxboard, but many products are wrapped individually or packed in bags or pouches. The trend for flexible packaging has gone toward synthetic material but, because paper is ecologically more acceptable, the trend might be turning.

3.7.1 Sack kraft

From a papermaker's point of view, sack kraft is of special interest. This paper, which combines all the common packaging functions mentioned above, is a highly developed, high-volume specialty paper. It combines two major requirements, toughness and porosity.

The paper needs to be strong enough to resist tough handling. It must be assumed that the sacks will be handled roughly, for instance, when unloaded from a lorry and thrown to the ground. This means that the paper must be able to absorb the shock or that the energy absorption capacity (called "work to burst") must be as high as possible. Work to

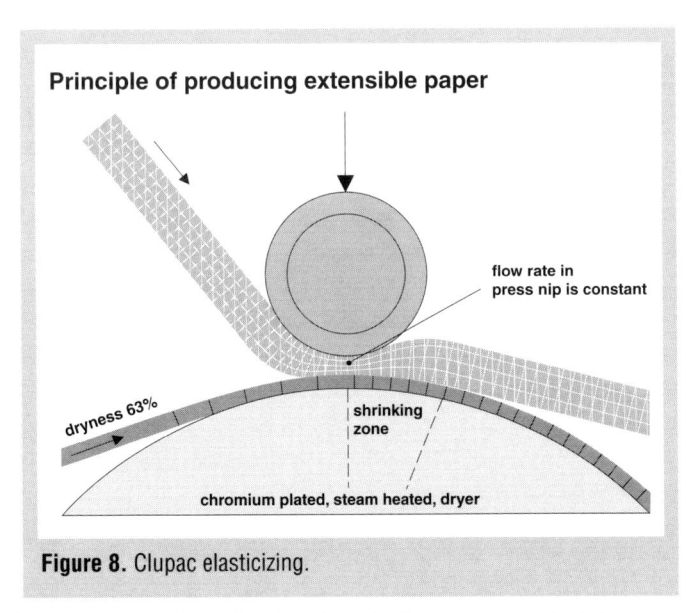

Figure 8. Clupac elasticizing.

burst is the elongation times the straining force. Further the tear strength must be good in order to avoid that minor damages spread further.

The second major requirement is that the sacks must be easily filled, which means the paper must be porous enough for not building up a counterpressure when the sack is filled.

These requirements lead to a manufacturing strategy, according to which the sack kraft paper is manufactured from medium to low refined long fiber kraft pulp, with refining at high and forming at a low consistency, granting good fiber-to-fiber bonding with a high porosity. Both bleached and unbleached fiber is used.

The paper is then dried with loose draws, allowing it to shrink in the dryer section of the paper machine. Thus microcreping is generated, which permits a high elongation before bursting. Typical values of elongation are as follows: 10% elongation in cross machine direction, and 4% elongation in machine direction.

In order to further improve the machine direction elongation the *clupac* method (see Fig. 8) is used, i.e., the paper is caused to microcrepe between a couple of very elastic cylinders.

Sack kraft is today produced on rather big, high efficiency paper machines, in a scale of 100 000 t/year or more. The paper sack is made of layers, the number of which is determined by the toughness required.

3.7.2 Flexible packaging

Flexible packaging papers, depending on the particular use, need combinations of printability, strength, and barrier characteristics (Fig. 9).

Twisting paper

Candy wrapping or twisting paper is a thin 30–40 g/m^2 kraft paper. It is mostly flexo or offset printed and slit into narrow coils, which are used in candy wrapping machines. The wrapping operation requires a good strength of the paper, which is made with a high MD/CD strength ratio, that is, with highly oriented fibers. This gives a good CD elongation, which allows the paper to follow the shape of the candy.

Twisting paper is mostly opaque, containing titanium dioxide as a pigment. It can also be pigmented on the surface in a size press, whereby titanium dioxide is dispersed in the size. Twisting paper for packaging is mostly supercalendered. For less demanding packaging purposes, there is a variety of similar papers, e.g., opaline, onion skin, and glassine.

Laminating papers

For uses, where an impermeable barrier is required, paper can be laminated or extrusion coated with polyethylene and other synthetic materials. Here again, a huge variety of different materials and applications exist.

Specialty papers

A typical lamination paper could be a coated and supercalendered thin kraft paper, 40–50 g/m² giving a good printability in rotogravure, and extrusion laminated with aluminum foil, which is hot-sealed to pouches for instant soup.

Figure 9. Flexible packages.

Grease proof, vegetable parchment

When a paper of pure cellulose is treated with sulfuric acid, the fibers swell, and a homogeneous film is formed. The result is vegetable parchment, which is produced in weights ranging from 40 to a maximum of 200 g/m². This traditional way of making a grease proof wrapping for butter and other fats has been replaced by synthetic materials or other methods.

A less expensive method is to refine the paper stock, creating a sheet with extremely low porosity. By supercalendering the paper, density is further improved. This method, yielding a paper called glassine, can be complemented by treating the paper in the size press – by starches, alginates, or CMC – filling the pores or chemically making the fiber surface fat repellent. Due to the difficult dewatering, glassine is usually produced in lower basis weights than vegetable parchment, mostly 30–50 g/m².

Fruit wrapping tissue

Citrus and also other fruits are sometimes wrapped in an individual anti-mold tissue. This is a 15–25 g/m² tissue that is treated in a size press or off machine with fungicide.

Wrapping papers

Wrapping papers comprise a very large variety of papers. They range from luxury wrapping tissues to coarse recovered paper-based sheets. Gift wrappings may be elaborately finished with fine printing, metallizing, or even laminating, but old newspapers may also serve as wrapping.

When talking about wrapping as a specialty, the most common grade would be an MG sheet of 40–70 g/m², without very special quality requirements. In the economy matrix, this would be a typical weather-vane grade of paper. The competitive value of high-grade wrappings mostly comes from converting and design, but this design can even be built into the very paper in the form of colors or sheet composition.

3.7.3 Label papers

Labeling is a simple method of marking nonprintable packaging materials like tins and bottles. In most cases, it is also easier to print a label than to print the entire packaging, for instance, when a product is distributed to different markets.

Depending on the use, the label papers are printed more or less sophisticatedly. The decorative effects can be enhanced by metallic surfaces made by lamination, vacuum metallizing, or even printing with metal pigments. Label papers are mostly one side or differentially coated (front for printing, back for gluing) in order to meet the demands of printing and further processing.

Beer and beverage label paper

Labels used on return bottles must be removable before reusing the bottle. Bottles are washed in washing machines using strongly caustic detergents. In order not to clog the washing machine, the labels must remain intact so that they can be removed from the washing machine by a coarse grid.

The body paper as well as the coating of the label must have a good wet and caustic strength, which is achieved by using wet-strength resins.

High-efficiency bottling lines (Fig. 10) set big demands on the label. The feeding must be smooth and secure, which requires a rough back side. The label must have a right degree of absorbency, stiffness, and dimensional stability in order to stick to the bottle during and after labeling. The problem is accentuated by the effect of the cold bottles condensing moisture from the environment.

The paper for beer and beverage labels is made of a mixture of short and long fiber bleached chemical pulp to a basis weight of about 70–75 g/m^2. Coating is mostly done by blade, and the paper is supercalendered.

There are also other varieties partially made with mechanical pulp.

Figure 10. Labeling unit running at a speed of 33 labels/second requires a particular set of characteristics of the paper.

Specialty papers

Pressure-sensitive labels

Pressure-sensitive labels are mostly used where the production volume is smaller or the variety is bigger than in beer and beverage packaging.

Typical applications are the cosmetics and pharmaceutical industries, wine and alcohols, and finally point of sale labeling.

The variety of uses implies a large number of demands on the label paper. The cosmetics and pharmaceutical industries mostly use luxury labels requiring the best possible printing. The paper then is high-quality and multiple- or cast-coated. Alternatively the label is varnished after printing.

Wine labels again aim for more rustic effects, and machine-finished, uncoated papers are used as well as coated ones. Here also semibleached furnishes and laid structured papers are sometimes used.

Vellum

The most common paper for pressure-sensitive labels is vellum, an uncoated woodfree paper of about 80 g/m^2 calendered to a satin-like finish. Vellum is used for pricing in POS (point of sale) and mailing labels in office use.

3.8 Specialty printing

Certain printed products require special characteristics, which may qualify them to be called specialties. Here a limit to specialty fine papers is difficult to ascertain.

Standard printing papers are much cheaper than the specialty papers; therefore, printers often use these even where a special paper would otherwise be required.

Map papers

Maps are used in different conditions and require a precise and detailed printing over a rather large sheet.

Sea chart board

Sea charts are used in different ways on huge merchant or cruising vessels as compared to coast guard, fishing, or leisure boats. They need to be tough and stiff enough to be handled in various circumstances, smooth enough for being put into piles of charts at the bridge of big ships, and resistant to folding in the small crafts. The paper must resist the circle and erasing for repeated navigation plotting, but also resist wind and water in the cockpit of small vessels and yachts.

Sea chart board is mostly around 170 g/m^2, made from a mixture of long and short fiber pulp. Some authorities even specify the use of textile fibers. The paper should be made with a good wet strength and the surface sized for good resistance. An MF paper finished to a good smoothness gives the right handling properties combined with good legibility.

CHAPTER 5

Terrain and city map paper

Terrain maps, used for orienteering, hiking, hunting, and military use, partly sets the same demands as sea charts. They must, however, be made of a thinner paper for handling in the terrain. A 80–100 g/m² strong MF paper, with good machine finishing and wet strength, is ideal for this use.

The pedestrian using a city map is in the same conditions as an orienteerer, and the map paper should be selected accordingly, i.e., it must be strong enough. Road maps are mostly used inside cars and do not need to be quite as strong as maps used outdoors.

Concert program, music paper

A printing paper can be made soft for not rattling when handled by using impregnation with a softener or particularly soft pulp.

3.9 Special fine papers

3.9.1 Security papers

Paper can be used as a means for avoiding forgery. The most typical counterfeiting objects are banknotes and personal identity documents, but also checks, bonds, share-certificates, and lottery or airline tickets are often printed on security paper with features difficult or impossible to imitate.

Watermarks are difficult to fake and easy to detect. Simpler watermarks can be made by the impression of a dandy roll or pressed into the paper on a wet press. More sophisticated half-tone marks are made with vat formers, where the forming wire is formed in a pattern corresponding to the watermark with the raised parts corresponding to light spots and the depressions corresponding to dark areas in the watermark.

The watermarks may be distributed as a pattern in the paper or positioned as a part of the design of the banknote, often as a portrait in an unprinted margin or shield.

Another traditional way to mark a security paper is to mingle it with colored fibers. The mingling fibers are long, thin textile fibers, which form a pattern of different colors embedded into the paper.

As techniques have developed, even more sophisticated security features are needed. The mingling can be done using optically active fibers so that the pattern becomes visible in ultraviolet light. Security threads with distinct patterns may be embedded into the paper when forming it. The thread may even be interlaced into a paper, so that it is partially embedded into the web and partly emerging on the surface.

Banknote papers are made to resist extended use. Among others, important strength characteristic is folding endurance. They are often made of cotton, linen, or hemp fibers and sized with gelatin. Production is typically on small paper machines.

3.9.2 Artists' papers

Artists' papers must be resistant to fading, neutral, and stable. The basis weight is mostly 200–400 g/m², and the finish varies from a coarse surface for painting to a very smooth surface for graphic work. Accordingly, the formation may be rough to very fine.

The furnish is bleached, low kappa chemical pulp, or even cotton or hemp fibers. For resistance to water, solvent, and erasing, gelatin or starch sizing is used.

In addition to the white painting and gravure papers, there are colored grades for special use. Artists' papers largely comprise a branding and distribution business with a great variety of types.

For scholastic use and sketching, normal uncoated 150–200 g/m^2 fine paper, preferably with a rather rough surface is used.

3.9.3 Tracing paper

The most used paper for technical drawings is transparent tracing paper. This paper is made out of highly refined chemical pulp in basis weights of 60–120 g/m^2. Due to the very slow stock, the paper is produced at low-speed, small fourdrinier machines. It is gelatin treated and supercalendered for good transparency and erasability.

Important characteristics are the uniform transparency for making diazo copies and a good degree of sizing for permitting drawing with china ink and erasing by knife. The paper is finished into customer rolls or sheets, in standard widths and sizes.

3.9.4 Plotting paper

CAD drawings are mostly reproduced on inkjet plotters, which require a suitable paper, with appropriate coating. Plotting paper is now quickly taking the market from tracing and diazo papers.

The coating must be fine-tuned to absorb the ink to an exact degree in order to avoid spreading of the lines. It is assumable that plotters and jet-printing ink will develop so that a standard grade of inkjet printing paper will be usable in most plotters.

3.9.5 Luxury stationery

When normal business stationery is too plain and ordinary, luxury brands are used. They are distinguished by certain features like texture, finish, and shade. They may be manufactured using prestigious fiber materials or even fake or real recycled paper fibers. The specialty with luxury stationery is image and design, with technical features staying in the background. Their sizing must, however, be suitable for fountain pen ink writing.

Luxury stationery can be produced by hand, on vat formers, or on fourdriniers. Hand-made papers mostly have a watermark of the manufacturer and may also be customized for the individual user by personal watermarks with initials, signature, or even a portrait. These papers are rarities and limited to a few users.

Machine-made papers are often made to imitate the hand-made ones. A laid structure similar to the bottom of old paper molds can easily be obtained by a dandy roll. Also the brand of the paper can be watermarked with a dandy roll and, with the aid of a reference point, the mark can even be positioned in the finished sheet.

Mingling with colored fibers is one common way of giving dignity and an antique touch to luxury papers. With a vat former, also half-tones and hand-torn papers similar to the hand-made papers can be made.

Luxury stationery is above all a design, branding, and distribution business.

3.10 Building papers

In buildings, paper is used to cover uneven surfaces and as a barrier to draught, covering permeable insulation materials.

3.10.1 Wallpaper

Wallpaper can carry a decor or be a painting ground. In the western world, demands on wallpapers change with fashion and level of sophistication in transforming. Wallpapers are printed in flexo, offset, or screen print, which all set different demands on the surface to be printed. A particular finish is obtained by screen printing with an expandable material, giving three-dimensional relief decor.

Certain wallpapers are made peelable, i.e., the top layer may be removed, leaving a wall prepared to receive a new wallpaper when redecorating the room. In this case, the paper is a duplex (two-layer) paper made with a loose ply bond strength. Such papers can be made on double-wire paper machines, couching the plies at a rather high dryness content and, if necessary, spraying a wax emulsion or similar hydrophobic agent between the layers when couching.

Wallpapers may thus be single-ply mechanical papers or duplex papers with a woodfree top layer and mechanical bottom. Washable grades are coated in the transformation phase, but may require a wet- and rubbing-resistant top layer.

Wallpapers are a specialty with a constant development for adapting to variations in fashion. The paper is typically from 80 to 120 g/m^2.

3.10.2 Barrier papers

Kraft papers or liner is used as a barrier or a wind shield in half timbered and similar constructions. For avoiding condensation and molding, the densest barrier comes toward the warmest surface. Therefore in cold climates, a polythene-coated kraft paper may be used to line the inner side of a wall, but the outer side must be permeable, either by porosity or through perforations.

3.11 Cigarette papers

Cigarettes are wrapped in a thin, 16–24 g/m^2 paper. The paper, in addition to being strong enough for working in cigarette wrapping machines, must have a porosity and ash content giving it the right glowing characteristics and leaving a good looking, white residue after burning.

Furthermore, the paper is often given a watermarked pattern with the dandy roll. It is made mainly of various textile fibers, linen, cotton, hemp, alpha grass, and special wood pulps. It is further treated with solutions of salts in the size press for controlling the burning properties.

Production is performed on small fourdrinier paper machines.

Specialty papers

Cork tip

A special cork imitation paper, made with barytha pigment, is used for many cigarette filters. This is also a thin paper with high wet strength and containing a yellowish brown pigment, which is printed to imitate cork.

3.12 Various functional papers

The selection above has illustrated the broadness of possible uses and particular demands on papers. There are hundreds of particular uses, and it would be beyond the scope of this book to try to list them in any completeness.

For instance, just think about papers and boards used for products such as:

- Surgical sterilization
- Wet refreshing towels
- Seeding pots
- Coffin lining
- Wax stencil
- Lamp shades
- Steel-plate interleaving
- Carbon copy.

All have their particular use and requirements and need specific development efforts, equipment, and know-how.

CHAPTER 5

Further reading

Göttsching, L. and Katz, C., Papier Lexikon, Deutscher Betriebswirte-Verlag, Gernsbach, 1999.

Conversion factors

To convert numerical values found in this book in the RECOMMENDED FORM, divide by the indicated number to obtain the values in CUSTOMARY UNITS. This table is an excerpt from TIS 0800-01 "Units of measurement and conversion factors." The complete document containing additional conversion factors and references to appropriate TAPPI Test Methods is available at no charge from TAPPI, Technology Park/Atlanta, P. O. Box 105113, Atlanta GA 30348-5113 (Telephone: +1 770 209-7303, 1-800-332-8686 in the United States, or 1-800-446-9431 in Canada).

Property	To convert values expressed in RECOMMENDED FORM	Divide by	To obtain values expressed In CUSTOMARY UNITS
Length	nanometers [nm]	0.1	angstroms [Å]
	micrometers [μm]	1	microns
	millimeters [mm]	0.0254	mils [mil or 0.001 in.]
	millimeters [mm]	25.4	inches [in.]
	meters [m]	0.3048	feet [ft]
	kilometers [km]	1.609	miles [mi]
Mass per unit area	grams per square meter [g/m^2]	3.7597	pounds per ream, 17 x 22 – 500
	grams per square meter [g/m^2]	1.4801	pounds per ream, 25 x 38 – 500
	grams per square meter [g/m^2]	1.4061	pounds per ream, 25 x 40 – 500
	grams per square meter [g/m^2]	4.8824	pounds per 1000 square feet [lb/1000 ft^2]
	grams per square meter [g/m^2]	1.6275	pounds per 3000 square feet [lb/3000 ft^2]
	grams per square meter [g/m^2]	1.6275	pounds per ream, 24 x 36 - 500
Pressure, stress, force per unit area	kilopascals [kPa]	100	bars [bar]
Speed	meters per second [m/s]	0.30480	feet per second [ft/s]
	millimeters per second [mm/s]	5.080	feet per minute [ft/min or fpm]
Thickness or caliper	micrometers [μm]	25.4	mils [mil] (or points or thousandths of an inch)
	millimeters [mm]	0.0254	mils [mil] (or 0.001 in.)
	millimeters [mm]	25.4	inches [in.]

Index

A
Abaca 111, 114
Abrasive base paper 115
Absorbent kraft 109
Acid wash 114
Adhesion 66, 71, 104
Air brush 120
Air filter 113
Anchoring 115–116
Appearance 43, 56, 64, 67, 82, 84, 105, 109–110
Artists' paper 126, 127
Automotive filter 113

B
Bag paper 114
Banknote 103, 126
Barrier 63, 102, 104, 107, 112, 122, 128
Barytha 129
Building paper 128
Book binding 71
Burning properties 128

C
Cake filtration 112
Caliber 117
Candy wrapping 122
Capacitor tissue 107
Carbon black 105, 108
Carbon copy 129
Carbonless copy 119–120
Casting paper 116
Chemical reactivity 102
Cigarette paper 116
Cleanliness 56–57, 63, 102, 105–106, 110–111
Clupac 121–122
Coffee filter 82
Coffin lining 129
Collating 120
Color changes 109
Color former 119–120

Colored paper 105
Commodity grade 101
Composite 109
Concert program 126
Conductive paper 108
Consumer packaging 64, 121
Continuous forms 120
Copy paper 39, 119–120
Cork tip 129
Corrugated board 64–67, 70, 91, 121
Corundum 110
cotton 79, 119, 126–128
  linters 119
Couching 128
Cover strips 118
Crepe 84, 88–89, 118
Curing 103, 111, 115, 117

D
Decor paper 109–110
Deionized water 107
Diapers 79, 97, 104, 118
Diazo 119, 127
Die cutting 117
Dielectric constant 106
Dimensional stability 29, 39–40, 104, 115, 120, 124
Dominating pore size 112
Dust holding capacity 111, 113

E
Electrical characteristics 105, 107
Electrical conductivity 102
Electrical insulation 102, 106
Electrical papers 106
Electroconductive resin 105
Elongation 103, 108, 116, 122
Embossing 57, 79, 83–84, 90–91, 118
Energy absorption 75, 122
Engine protection 111, 113
Entry barrier 102, 107
Envelope 104

Index

Erasing 125, 127
Eucalyptus 78, 111
Extrusion 119, 122–123

F
Fiber orientation 66, 116
Film press 117
Filter cartridge 102, 113
Filter paper 102, 111, 113
Filtration efficiency 111–112
Filtration mechanisms 112
Flash dried 111
Flexibility 75, 83, 116
Flexible packaging 104, 121–122
Folding boxboard 56, 58–61, 64, 70, 121
Formaldehyde 111
Formation 35, 60–62, 64, 67, 84, 103, 107, 109–110, 115, 119, 126
Fragmentation 102
Forming cylinder 108
Fruit wrapping 123
Fuel filter 114
Functional characteristics 56, 104, 107
Fungicide 123
Furniture papers 108

G
Gelatin 119, 126–127
Glassine 117, 122–123
Grease proof 123
Grinding belt 115

H
Half tone 126, 127
Hand finishing 116
Hand made paper 127
Heat sealing 114
Heavy duty filter 113
Hemp 111, 114, 126–128
High density laminate 108
Hunting cartridge 116

I
Imaging paper 118
Impregnation 105, 108, 110–111, 126
Inhibitor 120
Inorganic fibers 111

Internal tension 115, 120
Iron oxide 110

L
Label backing 104, 117
Label paper 117, 124–125
Laboratory filter 114
Laid structure 125, 127
Lamination, Laminate 82, 104, 108 110, 111, 117, 123, 124
Laminating kraft 111
Lamp shade 129
Laser printer 102
Life cycle 24, 102
Linen 126, 128
Luxury stationery 103, 105, 127

M
Machine glazing 46
Magazine paper 16, 28, 46, 102
Manual labels 118
Map paper 126
Masking tape 118
Mechanical strength 57, 103
Metallizing 123
Micro creping 103
Mingling 105, 126–127
Multi fourdrinier 64, 70, 71, 115
Multi-ply forms 119
Music paper 126

N
Nonwood fibers 111

O
Offset printing 28, 32, 34–35, 38, 58, 120
Onion skin 122
Opacity 24, 29, 32–35, 37, 48, 62, 67, 119
Opaline 122
Optical Character Reading 105
Overlay paper 109–110

P
Packaging 56–57, 59–64, 66, 82, 90–92, 95, 101, 104, 116, 121–122, 124–125
Particle board 108, 111
Permeability 112–113

Photographic paper 118
Pigment 27, 32, 34–35, 37–40, 43, 56, 60,
62, 108, 110, 118, 122, 129
Pinhole 107
Plotting 125, 127
Ply bond 115, 128
Pore size distribution 102, 104, 111–112
Porosity 33, 57, 66, 71, 97, 102, 104,
109–111, 114–115, 121–123, 128
Poster 45
Pouches 111, 121, 123
Power cable 106–107
Pre-impregnated foil 110,111
Pressure sensitive 117,125
Price elasticity 101
Price tag 120
Printability 24, 97, 122–123

R

Radioactivity 105
Refraction index 109
Release paper 104, 114, 118
Repellence 104
Return bottle 124
Rosin size 119

S

Sack kraft, sack 91, 121, 122
Scholastic 127
Screen printing 110, 128
Sea chart board 125
Security papers 103, 105, 126
Seeding pot 129
Self adhesive 116
Sensitivity matrix 115
Set gluing 120
Silicon, siliconizing 102, 104, 116–118
Size press 35, 46, 59–60, 62, 64, 66, 68,
104–105, 111, 115, 117, 119, 122–
123, 128
Spinning kraft 116
Star product 102, 115
Starch 37, 70, 115, 119, 127
Stationery 14, 39–40, 45–46, 103, 105,
120, 127
Steel plate interleaving 129
Stiffness 29, 32–35, 40, 49, 57, 59, 61–64,
71, 103, 111, 113, 124

Strength 24, 26, 29–30, 35–37, 40, 56–59,
62–64, 66, 68, 70–71, 77–78, 80, 82–
83, 85, 89, 95, 97, 102–104, 113–
116, 119, 122, 124–126, 128–129
Supercalender 33
Surface filtration 112–113
Surgical sterilization 104, 129
Synthetic fiber 114
Synthetic leather 118

T

Tea bag, filter 111,114
Tear strength 62, 66, 122
Telefax 120
Telephone cable 106
Terrain map 126
Thermal printing 120–121
Thickness 33, 57, 59, 82–84, 89, 102,
104, 107–108, 112–114, 117
Ticket 120, 121, 126
Titanium dioxide 62, 109, 122
Touch 105, 127
Toughness 103, 115, 121–122
Transparency 117–118, 127
Twisting 122

U

Underlay paper 109
Uniformity 103, 105, 109, 115

V

Vacuum cleaner 115
Vegetable parchment 123
Vellum 125
Volume filtration 112

W

Watermark 126–127
Wax emulsion 128
Wax stencil 129
Wear resistance 109
Wet grinding 115–116
Wet strength 64, 70, 77, 80, 103, 114,
116, 125–126, 129
Work to burst 122
Wrapping 102–103, 122–123, 128